Water Resources Engineering

Water Resources Engineering

Edited by
Sawyer Lopez

Larsen & Keller
www.larsen-keller.com

Water Resources Engineering
Edited by Sawyer Lopez
ISBN: 978-1-63549-291-0 (Hardback)

Larsen & Keller

Published by Larsen and Keller Education,
5 Penn Plaza,
19th Floor,
New York, NY 10001, USA

Cataloging-in-Publication Data

Water resources engineering / edited by Sawyer Lopez.
    p. cm.
Includes bibliographical references and index.
ISBN 978-1-63549-291-0
1. Hydraulic engineering. 2. Hydrology. 3. Waterworks. 4. Water resources development. I. Lopez, Sawyer.
TC145 .W38 2017
627--dc23

The publisher's policy is to use permanent paper from mills that operate a sustainable forestry policy. Furthermore, the publisher ensures that the text paper and cover boards used have met acceptable environmental accreditation standards.

Printed and bound in the United States of America.

For more information regarding Larsen and Keller Education and its products, please visit the publisher's website www.larsen-keller.com

Table of Contents

Preface

This book provides comprehensive insights about water resources. It attempts to understand the multiple branches that fall under this discipline and how such concepts have uses in our environment. Water resources refer to the naturally occurring water bodies, which are useful in industrial, agricultural, household and environmental activities. This book picks up individual branches and explains their need and contribution in the context of a growing economy. The aim of this text is to provide the students the basic idea of this field. It is a compilation of chapters that discuss the most vital concepts in the field of water resources and its necessity and management. Coherent flow of topics, student-friendly language and extensive use of examples make this textbook an invaluable source of knowledge.

A short introduction to every chapter is written below to provide an overview of the content of the book:

Chapter 1 - The sources of water that are useful or can be potentially useful are known as water resources. Humans majorly use fresh water, which is only 3 percent of the Earth's water. The following text is an overview of the subject matter incorporating all the major aspects of water resources; **Chapter 2 -** Ground water is an important water resource. Ground water is present beneath the surface of the Earth and is naturally renewed. This chapter also elucidates seawater and surface water and provides a plethora of interdisciplinary topics for a better comprehension of water resources; **Chapter 3 -** The chapter illustrates and demonstrates the study of fresh water. Fresh water is naturally occurring water on earth's surface in ice sheets, ice caps, rivers and streams, and groundwater. The following content also gives a profound content on desalination. The topics discussed in the chapter are of great importance to broaden the existing knowledge on water resources; **Chapter 4 -** To understand storm water elaborately, the following text gives an insight on storm water harvesting. Storm water harvesting is the collection, accumulation, treatment or purification, and storing of storm water for its eventual reuse. The chapter serves as a source to understand the major categories related to storm water; **Chapter 5 -** Integrated water resources management is defined as a process which promotes the coordinated development and management of water, land and related resources. The program takes into account both urban and rural areas for water resources development. The content gives an overview of the subject matter incorporating all the major aspects of water integration; **Chapter 6 -** The major concepts of irrigation are discussed in this chapter. Deficit irrigation, flexible barge, hippo water roller and peak water are important topics related to irrigation strategies. The chapter strategically encompasses and incorporates the major components and key concepts of irrigation strategies, providing a complete understanding; **Chapter 7 -** The policies and strategies concerned with water conservation in order to protect the hydrosphere and to meet future demands is water conservation. This text provides a plethora of interdisciplinary topics for better comprehension of water conservation; **Chapter 8 -** Water conflict is a term describing a conflict between countries, states or groups over water resources. Water has historically been a source of tension and a factor of conflicts; this chapter elaborates this aspect on water. The following chapter will provide an integrated understanding of water resources; **Chapter 9 -** Planning is an important component of any field of study. Shared vision planning seeks to be an update on existing water resource management systems. Decisions are made through a process of "informed consent. "The content serves as a source to understand the process of shared vision planning; **Chapter 10** - Water resources can best be understood with the major topics listed in the following chapter. The laws, exercise and policy related to water resources are dealt with in this chapter. Water quality law, water resources law, water rights and xerochore are some of the topics elaborated in the concerned text.

I extend my sincere thanks to the publisher for considering me worthy of this task. Finally, I thank my family for being a source of support and help.

Editor

Introduction to Water Resources

The sources of water that are useful or can be potentially useful are known as water resources. Humans majorly use fresh water, which is only 3 percent of the Earth's water. The following text is an overview of the subject matter incorporating all the major aspects of water resources.

Water Resources

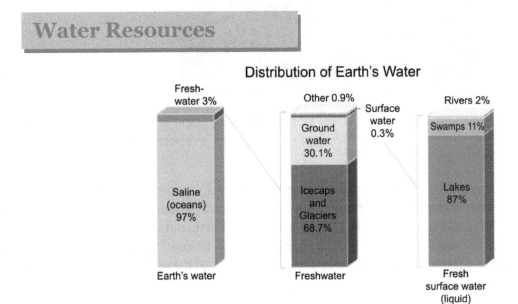

A graphical distribution of the locations of water on Earth. Only 3% of the Earth's water is fresh water. Most of it is in icecaps and glaciers (69%) and groundwater (30%), while all lakes, rivers and swamps combined only account for a small fraction (0.3%) of the Earth's total freshwater reserves.

Water resources are sources of water that are useful or potentially useful. Uses of water include agricultural, industrial, household, recreational and environmental activities. The majority of human uses require fresh water.

97 % of the water on the Earth is salt water and only three percent is fresh water; slightly over two thirds of this is frozen in glaciers and polar ice caps. The remaining unfrozen freshwater is found mainly as groundwater, with only a small fraction present above ground or in the air.

Fresh water is a renewable resource, yet the world's supply of groundwater is steadily decreasing, with depletion occurring most prominently in Asia and North America, although it is still unclear how much natural renewal balances this usage, and whether ecosystems are threatened. The framework for allocating water resources to water users (where such a framework exists) is known as water rights.

Sources of Fresh Water

Surface Water

Lake Chungará and Parinacota volcano in northern Chile

Surface water is water in a river, lake or fresh water wetland. Surface water is naturally replenished by precipitation and naturally lost through discharge to the oceans, evaporation, evapotranspiration and groundwater recharge.

Although the only natural input to any surface water system is precipitation within its watershed, the total quantity of water in that system at any given time is also dependent on many other factors. These factors include storage capacity in lakes, wetlands and artificial reservoirs, the permeability of the soil beneath these storage bodies, the runoff characteristics of the land in the watershed, the timing of the precipitation and local evaporation rates. All of these factors also affect the proportions of water loss.

Human activities can have a large and sometimes devastating impact on these factors. Humans often increase storage capacity by constructing reservoirs and decrease it by draining wetlands. Humans often increase runoff quantities and velocities by paving areas and channelizing stream flow.

The total quantity of water available at any given time is an important consideration. Some human water users have an intermittent need for water. For example, many farms require large quantities of water in the spring, and no water at all in the winter. To supply such a farm with water, a surface water system may require a large storage capacity to collect water throughout the year and release it in a short period of time. Other users have a continuous need for water, such as a power plant that requires water for cooling. To supply such a power plant with water, a surface water system only needs enough storage capacity to fill in when average stream flow is below the power plant's need.

Nevertheless, over the long term the average rate of precipitation within a watershed is the upper bound for average consumption of natural surface water from that watershed.

Natural surface water can be augmented by importing surface water from another watershed through a canal or pipeline. It can also be artificially augmented from any of the other sources listed here, however in practice the quantities are negligible. Humans can also cause surface water to be "lost" (i.e. become unusable) through pollution.

Brazil is the country estimated to have the largest supply of fresh water in the world, followed by Russia and Canada.

Panorama of a natural wetland (Sinclair Wetlands, New Zealand)

Under River Flow

Throughout the course of a river, the total volume of water transported downstream will often be a combination of the visible free water flow together with a substantial contribution flowing through rocks and sediments that underlie the river and its floodplain called the hyporheic zone. For many rivers in large valleys, this unseen component of flow may greatly exceed the visible flow. The hyporheic zone often forms a dynamic interface between surface water and groundwater from aquifers, exchanging flow between rivers and aquifers that may be fully charged or depleted. This is especially significant in karst areas where pot-holes and underground rivers are common.

Groundwater

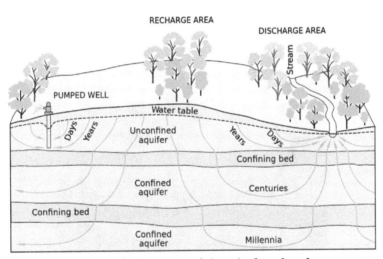

Relative groundwater travel times in the subsurface

Groundwater is fresh water located in the subsurface pore space of soil and rocks. It is also water that is flowing within aquifers below the water table. Sometimes it is useful to make a distinction between groundwater that is closely associated with surface water and deep groundwater in an aquifer (sometimes called "fossil water").

Groundwater can be thought of in the same terms as surface water: inputs, outputs and storage. The critical difference is that due to its slow rate of turnover, groundwater storage is generally much larger (in volume) compared to inputs than it is for surface water. This difference makes it easy for humans to use groundwater unsustainably for a long time without severe consequences.

Nevertheless, over the long term the average rate of seepage above a groundwater source is the upper bound for average consumption of water from that source.

A shipot is a common water source in Central Ukrainian villages

The natural input to groundwater is seepage from surface water. The natural outputs from groundwater are springs and seepage to the oceans.

If the surface water source is also subject to substantial evaporation, a groundwater source may become saline. This situation can occur naturally under endorheic bodies of water, or artificially under irrigated farmland. In coastal areas, human use of a groundwater source may cause the direction of seepage to ocean to reverse which can also cause soil salinization. Humans can also cause groundwater to be "lost" (i.e. become unusable) through pollution. Humans can increase the input to a groundwater source by building reservoirs or detention ponds.

Frozen Water

Iceberg near Newfoundland

Several schemes have been proposed to make use of icebergs as a water source, however to date this has only been done for research purposes. Glacier runoff is considered to be surface water.

The Himalayas, which are often called "The Roof of the World", contain some of the most extensive and rough high altitude areas on Earth as well as the greatest area of glaciers and permafrost outside of the poles. Ten of Asia's largest rivers flow from there, and more than a billion people's livelihoods depend on them. To complicate matters, temperatures there are rising more rapidly than the global average. In Nepal, the temperature has risen by 0.6 degrees Celsius over the last decade, whereas globally, the Earth has warmed approximately 0.7 degrees Celsius over the last hundred years.

Desalination

Desalination is an artificial process by which saline water (generally sea water) is converted to fresh water. The most common desalination processes are distillation and reverse osmosis. Desalination is currently expensive compared to most alternative sources of water, and only a very small fraction of total human use is satisfied by desalination. It is only economically practical for high-valued uses (such as household and industrial uses) in arid areas. The most extensive use is in the Persian Gulf.

Water Uses

Agricultural

It is estimated that 70% of worldwide water is used for irrigation, with 15-35% of irrigation withdrawals being unsustainable. It takes around 2,000 - 3,000 litres of water to produce enough food to satisfy one person's daily dietary need. This is a considerable amount, when compared to that required for drinking, which is between two and five litres. To produce food for the now over 7 billion people who inhabit the planet today requires the water that would fill a canal ten metres deep, 100 metres wide and 2100 kilometres long.

Increasing Water Scarcity

Around fifty years ago, the common perception was that water was an infinite resource. At that time, there were fewer than half the current number of people on the planet. People were not as wealthy as today, consumed fewer calories and ate less meat, so less water was needed to produce their food. They required a third of the volume of water we presently take from rivers. Today, the competition for water resources is much more intense. This is because there are now seven billion people on the planet, their consumption of water-thirsty meat and vegetables is rising, and there is increasing competition for water from industry, urbanisation biofuel crops, and water reliant food items. In the future, even more water will be needed to produce food because the Earth's population is forecast to rise to 9 billion by 2050. An additional 2.5 or 3 billion people, choosing to eat fewer cereals and more meat and vegetables could add an additional five million kilometres to the virtual canal mentioned above.

An assessment of water management in agriculture sector was conducted in 2007 by the International Water Management Institute in Sri Lanka to see if the world had sufficient water to provide food for its growing population. It assessed the current availability of water for agriculture on a global scale and mapped out locations suffering from water scarcity. It found that a fifth of the world's people, more than 1.2 billion, live in areas of physical water scarcity, where there is not enough water to meet all demands. One third of the worlds population does not have access to clean drinking water, which is more than 2.3 billion people. A further 1.6 billion people live in areas experiencing economic water scarcity, where the lack of investment in water or insufficient human capacity make it impossible for authorities to satisfy the demand for water. The report found that it would be possible to produce the food required in future, but that continuation of today's food production and environmental trends would lead to crises in many parts of the world. To avoid a global water crisis, farmers will have to strive to increase productivity to meet growing demands for food, while industry and cities find ways to use water more efficiently.

In some areas of the world, irrigation is necessary to grow any crop at all, in other areas it permits more profitable crops to be grown or enhances crop yield. Various irrigation methods involve different trade-offs between crop yield, water consumption and capital cost of equipment and structures. Irrigation methods such as furrow and overhead sprinkler irrigation are usually less expensive but are also typically less efficient, because much of the water evaporates, runs off or drains below the root zone. Other irrigation methods considered to be more efficient include drip or trickle irrigation, surge irrigation, and some types of sprinkler systems where the sprinklers are operated near ground level. These types of systems, while more expensive, usually offer greater potential to minimize runoff, drainage and evaporation. Any system that is improperly managed can be wasteful, all methods have the potential for high efficiencies under suitable conditions, appropriate irrigation timing and management. Some issues that are often insufficiently considered are salinization of groundwater and contaminant accumulation leading to water quality declines.

As global populations grow, and as demand for food increases in a world with a fixed water supply, there are efforts under way to learn how to produce more food with less water, through improvements in irrigation methods and technologies, agricultural water management, crop types, and water monitoring. Aquaculture is a small but growing agricultural use of water. Freshwater commercial fisheries may also be considered as agricultural uses of water, but have generally been assigned a lower priority than irrigation.

Industrial

A power plant in Poland

It is estimated that 22% of worldwide water is used in industry. Major industrial users include hydroelectric dams, thermoelectric power plants, which use water for cooling, ore and oil refineries, which use water in chemical processes, and manufacturing plants, which use water as a solvent. Water withdrawal can be very high for certain industries, but consumption is generally much lower than that of agriculture.

Water is used in renewable power generation. Hydroelectric power derives energy from the force of water flowing downhill, driving a turbine connected to a generator. This hydroelectricity is a low-cost, non-polluting, renewable energy source. Significantly, hydroelectric power can also be used for load following unlike most renewable energy sources which are intermittent. Ultimately, the energy in a hydroelectric powerplant is supplied by the sun. Heat from the sun evaporates water, which condenses as rain in higher altitudes and flows downhill. Pumped-storage hydroelectric plants also exist, which use grid electricity to pump water uphill when demand is low, and use the stored water to produce electricity when demand is high.

Hydroelectric power plants generally require the creation of a large artificial laWke. Evaporation from this lake is higher than evaporation from a river due to the larger surface area exposed to the elements, resulting in much higher water consumption. The process of driving water through the turbine and tunnels or pipes also briefly removes this water from the natural environment, creating water withdrawal. The impact of this withdrawal on wildlife varies greatly depending on the design of the powerplant.

Pressurized water is used in water blasting and water jet cutters. Also, very high pressure water guns are used for precise cutting. It works very well, is relatively safe, and is not harmful to the environment. It is also used in the cooling of machinery to prevent overheating, or prevent saw blades from overheating. This is generally a very small source of water consumption relative to other uses.

Water is also used in many large scale industrial processes, such as thermoelectric power production, oil refining, fertilizer production and other chemical plant use, and natural gas extraction from shale rock. Discharge of untreated water from industrial uses is pollution. Pollution includes discharged solutes (chemical pollution) and increased water temperature (thermal pollution). Industry requires pure water for many applications and utilizes a variety of purification techniques both in water supply and discharge. Most of this pure water is generated on site, either from natural freshwater or from municipal grey water. Industrial consumption of water is generally much lower than withdrawal, due to laws requiring industrial grey water to be treated and returned to the environment. Thermoelectric powerplants using cooling towers have high consumption, nearly equal to their withdrawal, as most of the withdrawn water is evaporated as part of the cooling process. The withdrawal, however, is lower than in once-through cooling systems.

Domestic

Drinking water

It is estimated that 8% of worldwide water use is for domestic purposes. These include drinking water, bathing, cooking, toilet flushing, cleaning, laundry and gardening. Basic domestic water requirements have been estimated by Peter Gleick at around 50 liters per person per day, excluding water for gardens. Drinking water is water that is of sufficiently high quality so that it can be consumed or used without risk of immediate or long term harm. Such water is commonly called potable water. In most developed countries, the water supplied to domestic, commerce and industry is all of drinking water standard even though only a very small proportion is actually consumed or used in food preparation.

Recreation

Whitewater rapids

Recreational water use is usually a very small but growing percentage of total water use. Recreational water use is mostly tied to reservoirs. If a reservoir is kept fuller than it would otherwise be for recreation, then the water retained could be categorized as recreational usage. Release of water from a few reservoirs is also timed to enhance whitewater boating, which also could be considered a recreational usage. Other examples are anglers, water skiers, nature enthusiasts and swimmers.

Recreational usage is usually non-consumptive. Golf courses are often targeted as using excessive amounts of water, especially in drier regions. It is, however, unclear whether recreational irrigation (which would include private gardens) has a noticeable effect on water resources. This is largely due to the unavailability of reliable data. Additionally, many golf courses utilize either primarily or exclusively treated effluent water, which has little impact on potable water availability.

Some governments, including the Californian Government, have labelled golf course usage as agricultural in order to deflect environmentalists' charges of wasting water. However, using the above figures as a basis, the actual statistical effect of this reassignment is close to zero. In Arizona, an organized lobby has been established in the form of the Golf Industry Association, a group focused on educating the public on how golf impacts the environment.

Recreational usage may reduce the availability of water for other users at specific times and places. For example, water retained in a reservoir to allow boating in the late summer is not available to farmers during the spring planting season. Water released for whitewater rafting may not be available for hydroelectric generation during the time of peak electrical demand.

Environmental

Explicit environment water use is also a very small but growing percentage of total water use. Environmental water may include water stored in impoundments and released for environmental purposes (held environmental water), but more often is water retained in waterways through regulatory limits of abstraction. Environmental water usage includes watering of natural or artificial wetlands, artificial lakes intended to create wildlife habitat, fish ladders, and water releases from reservoirs timed to help fish spawn, or to restore more natural flow regimes

Like recreational usage, environmental usage is non-consumptive but may reduce the availability of water for other users at specific times and places. For example, water release from a reservoir to help fish spawn may not be available to farms upstream, and water retained in a river to maintain waterway health would not be available to water abstractors downstream.

Water Stress

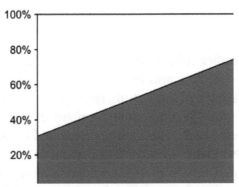

Estimate of the share of people in developing countries with access to drinking water 1970–2000

The concept of water stress is relatively simple: According to the World Business Council for Sustainable Development, it applies to situations where there is not enough water for all uses, whether agricultural, industrial or domestic. Defining thresholds for stress in terms of available water per capita is more complex, however, entailing assumptions about water use and its efficiency. Nevertheless, it has been proposed that when annual per capita renewable freshwater availability is less than 1,700 cubic meters, countries begin to experience periodic or regular water stress. Below 1,000 cubic meters, water scarcity begins to hamper economic development and human health and well-being.

Population Growth

In 2000, the world population was 6.2 billion. The UN estimates that by 2050 there will be an additional 3.5 billion people with most of the growth in developing countries that already suffer water stress. Thus, water demand will increase unless there are corresponding increases in water conservation and recycling of this vital resource. In building on the data presented here by the UN, the World Bank goes on to explain that access to water for producing food will be one of the main challenges in the decades to come. Access to water will need to be balanced with the importance of managing water itself in a sustainable way while taking into account the impact of climate change, and other environmental and social variables.

Expansion of Business Activity

Business activity ranging from industrialization to services such as tourism and entertainment continues to expand rapidly. This expansion requires increased water services including both supply and sanitation, which can lead to more pressure on water resources and natural ecosystem.

Rapid Urbanization

The trend towards urbanization is accelerating. Small private wells and septic tanks that work well in low-density communities are not feasible within high-density urban areas. Urbanization requires significant investment in water infrastructure in order to deliver water to individuals and to process the concentrations of wastewater – both from individuals and from business. These polluted and contaminated waters must be treated or they pose unacceptable public health risks.

In 60% of European cities with more than 100,000 people, groundwater is being used at a faster rate than it can be replenished. Even if some water remains available, it costs increasingly more to capture it.

Climate Change

Climate change could have significant impacts on water resources around the world because of the close connections between the climate and hydrological cycle. Rising temperatures will increase evaporation and lead to increases in precipitation, though there will be regional variations in rainfall. Both droughts and floods may become more frequent in different regions at different times, and dramatic changes in snowfall and snow melt are expected in mountainous areas. Higher temperatures will also affect water quality in ways that are not well understood. Possible impacts include increased eutrophication. Climate change could also mean an increase in demand for farm irrigation, garden sprinklers, and perhaps even swimming pools. There is now ample evidence that increased hydrologic variability and change in climate has and will continue have a profound impact on the water sector through the hydrologic cycle, water availability, water demand, and water allocation at the global, regional, basin, and local levels.

Depletion of Aquifers

Due to the expanding human population, competition for water is growing such that many of the world's major aquifers are becoming depleted. This is due both for direct human consumption as well as agricultural irrigation by groundwater. Millions of pumps of all sizes are currently extracting groundwater throughout the world. Irrigation in dry areas such as northern China, Nepal and India is supplied by groundwater, and is being extracted at an unsustainable rate. Cities that have experienced aquifer drops between 10 and 50 meters include Mexico City, Bangkok, Beijing, Madras and Shanghai.

Pollution and Water Protection

Water pollution is one of the main concerns of the world today. The governments of numerous countries have striven to find solutions to reduce this problem. Many pollutants threaten water supplies, but the most widespread, especially in developing countries, is the discharge of raw sewage into natural waters; this method of sewage disposal is the most common method in underdeveloped countries, but also is prevalent in quasi-developed countries such as China, India, Nepal and Iran. Sewage, sludge, garbage, and even toxic pollutants are all dumped into the water. Even if sewage is treated, problems still arise. Treated sewage forms sludge, which may be placed in landfills, spread out on land, incinerated or dumped at sea. In addition to sewage, nonpoint source pollution such as agricultural runoff is a significant source of pollution in some parts of the world, along with urban stormwater runoff and chemical wastes dumped by industries and governments.

Polluted water

Water and Conflicts

Competition for water has widely increased, and it has become more difficult to conciliate the necessities for water supply for human consumption, food production, ecosystems and other uses. Water administration is frequently involved in contradictory and complex problems. Approximately 10% of the worldwide annual runoff is used for human necessities. Several areas of the world are flooded, while others have such low precipitations that human life is almost impossible. As population and development increase, raising water demand, the possibility of problems inside a certain country or region increases, as it happens with others outside the region.

Over the past 25 years, politicians, academics and journalists have frequently predicted that disputes over water would be a source of future wars. Commonly cited quotes include: that of former Egyptian Foreign Minister and former Secretary-General of the United Nations Boutrous Ghali, who forecast, "The next war in the Middle East will be fought over water, not politics"; his successor at the UN, Kofi Annan, who in 2001 said, "Fierce competition for fresh water may well become a source of conflict and wars in the future," and the former Vice President of the World Bank, Ismail Serageldin, who said the wars of the next century will be over water unless significant changes in governance occurred. The water wars hypothesis had its roots in earlier research carried out on a small number of transboundary rivers such as the Indus, Jordan and Nile. These particular rivers became the focus because they had experienced water-related disputes. Specific events cited as evidence include Israel's bombing of Syria's attempts to divert the Jordan's headwaters, and military threats by Egypt against any country building dams in the upstream waters of the Nile. However, while some links made between conflict and water were valid, they did not necessarily represent the norm.

The only known example of an actual inter-state conflict over water took place between 2500 and 2350 BC between the Sumerian states of Lagash and Umma. Water stress has most often led to conflicts at local and regional levels. Tensions arise most often within national borders, in the downstream areas of distressed river basins. Areas such as the lower regions of China's Yellow River or the Chao Phraya River in Thailand, for example, have already been experiencing water stress for several years. Water stress can also exacerbate conflicts and political tensions which are not directly caused by water. Gradual reductions over time in the quality and/or quantity of fresh water can add to the instability of a region by depleting the health of a population, obstructing economic development, and exacerbating larger conflicts.

Shared Water Resources can Promote Collaboration

Water resources that span international boundaries are more likely to be a source of collaboration and cooperation than war. Scientists working at the International Water Management Institute have been investigating the evidence behind water war predictions. Their findings show that, while it is true there has been conflict related to water in a handful of international basins, in the rest of the world's approximately 300 shared basins the record has been largely positive. This is exemplified by the hundreds of treaties in place guiding equitable water use between nations sharing water resources. The institutions created by these agreements can, in fact, be one of the most important factors in ensuring cooperation rather than conflict.

The International Union for the Conservation of Nature (IUCN) published the book *Share: Managing water across boundaries*. One chapter covers the functions of trans-boundary institutions and how they can be designed to promote cooperation, overcome initial disputes and find ways of coping with the uncertainty created by climate change. It also covers how the effectiveness of such institutions can be monitored.

Water Shortages

In 2025, water shortages will be more prevalent among poorer countries where resources are limited and population growth is rapid, such as the Middle East, Africa, and parts of Asia. By 2025, large urban and peri-urban areas will require new infrastructure to provide safe water and adequate sanitation. This suggests growing conflicts with agricultural water users, who currently consume the majority of the water used by humans.

Generally speaking the more developed countries of North America, Europe and Russia will not see a serious threat to water supply by the year 2025, not only because of their relative wealth, but more importantly their populations will be better aligned with available water resources. North Africa, the Middle East, South Africa and northern China will face very severe water shortages due to physical scarcity and a condition of overpopulation relative to their carrying capacity with respect to water supply. Most of South America, Sub-Saharan Africa, Southern China and India will face water supply shortages by 2025; for these latter regions the causes of scarcity will be economic constraints to developing safe drinking water, as well as excessive population growth.

Economic Considerations

Water supply and sanitation require a huge amount of capital investment in infrastructure such as pipe networks, pumping stations and water treatment works. It is estimated that Organisation for Economic Co-operation and Development (OECD) nations need to invest at least USD 200 billion per year to replace aging water infrastructure to guarantee supply, reduce leakage rates and protect water quality.

International attention has focused upon the needs of the developing countries. To meet the Millennium Development Goals targets of halving the proportion of the population lacking access to safe drinking water and basic sanitation by 2015, current annual investment on the order of USD 10 to USD 15 billion would need to be roughly doubled. This does not include investments required for the maintenance of existing infrastructure.

Once infrastructure is in place, operating water supply and sanitation systems entails significant ongoing costs to cover personnel, energy, chemicals, maintenance and other expenses. The sources of money to meet these capital and operational costs are essentially either user fees, public funds or some combination of the two. An increasing dimension to consider is the flexibility of the water supply system.

Water Resource Management

Water resource management is the activity of planning, developing, distributing and managing the optimum use of water resources. It is a sub-set of water cycle management. Ideally, water resource management planning has regard to all the competing demands for water and seeks to allocate water on an equitable basis to satisfy all uses and demands. As with other resource management, this is rarely possible in practice.

Overview

Water is an essential resource for all life on the planet. Of the water resources on Earth only three percent of it is fresh and two-thirds of the freshwater is locked up in ice caps and glaciers. Of the remaining one percent, a fifth is in remote, inaccessible areas and much seasonal rainfall in monsoonal deluges and floods cannot easily be used. As time advances, water is becoming scarcer and having access to clean, safe, drinking water is limited among countries. At present only about 0.08 percent of all the world's fresh water is exploited by mankind in ever increasing demand for sanitation, drinking, manufacturing, leisure and agriculture. Due to the small percentage of water remaining, optimizing the fresh water we have left from natural resources has been a continuous difficulty in several locations worldwide.

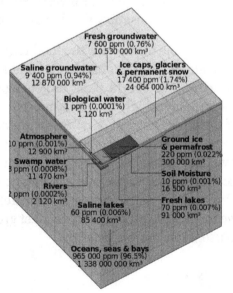

Visualisation of the distribution (by volume) of water on Earth. Each tiny cube (such as the one representing biological water) corresponds to approximately 1000 cubic km of water, with a mass of approximately 1 trillion tonnes (2000 times that of the Great Pyramid of Giza or 5 times that of Lake Kariba, arguably the heaviest man-made object). The entire block comprises 1 million tiny cubes.

Much efforts in water resource management is directed at optimizing the use of water and in minimizing the environmental impact of water use on the natural environment. The observation of water as an integral part of the ecosystem is based on integrated water resource management, where the quantity and quality of the ecosystem help to determine the nature of the natural resources.

Successful management of any resources requires accurate knowledge of the resource available, the uses to which it may be put, the competing demands for the resource, measures to and processes to evaluate the significance and worth of competing demands and mechanisms to translate policy decisions into actions on the ground.

For water as a resource this is particularly difficult since sources of water can cross many national boundaries and the uses of water include many that are difficult to assign financial value to and may also be difficult to manage in conventional terms. Examples include rare species or ecosystems or the very long term value of ancient ground water reserves.

Agriculture

Agriculture is the largest user of the world's freshwater resources, consuming 70 percent. As the world population rises it consumes more food (currently exceeding 6%, it is expected to reach 9% by 2050), the industries and urban developments expand, and the emerging biofuel crops trade also demands a share of freshwater resources, water scarcity is becoming an important issue. An assessment of water resource management in agriculture was conducted in 2007 by the International Water Management Institute in Sri Lanka to see if the world had sufficient water to provide food for its growing population or not . It assessed the current availability of water for agriculture on a global scale and mapped out locations suffering from water scarcity. It found that a fifth of the world's people, more than 1.2 billion, live in areas of physical water scarcity, where there is not enough water to meet all their demands. A further 1.6 billion people live in areas experiencing economic water scarcity, where the lack of investment in water or insufficient human capacity make it impossible for authorities to satisfy the demand for water.

The report found that it would be possible to produce the food required in future, but that continuation of today's food production and environmental trends would lead to crises in many parts of the world. Regarding food production, the World Bank targets agricultural food production and water resource management as an increasingly global issue that is fostering an important and growing debate. The authors of the book *Out of Water: From abundance to Scarcity and How to Solve the World's Water Problems*, which laid down a six-point plan for solving the world's water problems. These are: 1) Improve data related to water; 2) Treasure the environment; 3) Reform water governance; 4) Revitalize agricultural water use; 5) Manage urban and industrial demand; and 6) Empower the poor and women in water resource management. To avoid a global water crisis, farmers will have to strive to increase productivity to meet growing demands for food, while industry and cities find ways to use water more efficiently.

Managing Water in Urban Settings

As the carrying capacity of the Earth increases greatly due to technological advances, urbanization in modern times occurs because of economic opportunity. This rapid urbanization happens worldwide but mostly in new rising economies and developing countries. Cities in Africa and Asia are growing fastest with 28 out of 39 megacities (a city or urban area with more than 10 million inhabitants) worldwide in these developing nations. The number of megacities will continue to rise reaching approximately 50 in 2025. With developing economies water scarcity is a very common and very prevalent issue. Global freshwater resources dwindle in the eastern hemisphere either than at the poles, and with the majority of urban development millions live with insufficient fresh water. This is caused by polluted freshwater resources, overexploited groundwater resources, insufficient harvesting capacities in the surrounding rural areas, poorly constructed and maintained water supply systems, high amount of informal water use and insufficient technical and water management capacities.

In the areas surrounding urban centres, agriculture must compete with industry and municipal users for safe water supplies, while traditional water sources are becoming polluted with urban wastewater. As cities offer the best opportunities for selling produce, farmers often have no alternative to using polluted water to irrigate their crops. Depending on how developed a city's wastewater treatment is, there can be significant health hazards related to the use of this water. Wastewater from cities can contain a mixture of pollutants. There is usually wastewater from kitchens and toilets along with rainwater runoff. This means that the water usually contains excessive levels of nutrients and salts, as well as a wide range of pathogens. Heavy metals may also be present, along with traces of antibiotics and endocrine disruptors, such as oestrogens.

Developing world countries tend to have the lowest levels of wastewater treatment. Often, the water that farmers use for irrigating crops is contaminated with pathogens from sewage. The pathogens of most concern are bacteria, viruses and parasitic worms, which directly affect farmers' health and indirectly affect consumers if they eat the contaminated crops. Common illnesses include diarrhoea, which kills 1.1 million people annually and is the second most common cause of infant deaths. Many cholera outbreaks are also related to the reuse of poorly treated wastewater. Actions that reduce or remove contamination, therefore, have the potential to save a large number of lives and improve livelihoods. Scientists have been working to find ways to reduce contamination of food using a method called the 'multiple-barrier approach'.

This involves analysing the food production process from growing crops to selling them in markets and eating them, then considering where it might be possible to create a barrier against contamination. Barriers include: introducing safer irrigation practices; promoting on-farm wastewater treatment; taking actions that cause pathogens to die off; and effectively washing crops after harvest in markets and restaurants.

Future of Water Resources

One of the biggest concerns for our water-based resources in the future is the sustainability of the current and even future water resource allocation. As water becomes more scarce, the importance of how it is managed grows vastly. Finding a balance between what is needed by humans and what is needed in the environment is an important step in the sustainability of water resources. Attempts to create sustainable freshwater systems have been seen on a national level in countries such as Australia, and such commitment to the environment could set a model for the rest of the world.

The field of water resources management will have to continue to adapt to the current and future issues facing the allocation of water. With the growing uncertainties of global climate change and the long term impacts of management actions,the decision-making will be even more difficult. It is likely that ongoing climate change will lead to situations that have not been encountered. As a result, new management strategies will have to be implemented in order to avoid setbacks in the allocation of water resources.

References

- Mund, Jan-Peter. "Capacities for Megacities coping with water scarcity" (PDF). UN-Water Decade Programme on Capacity Development. Retrieved 2014-02-17.

- Gleeson, Tom; Wada, Yoshihide; Bierkens, Marc F. P.; van Beek, Ludovicus P. H. (9 August 2012). "Water balance of global aquifers revealed by groundwater footprint". Nature. 488 (7410): 197–200. doi:10.1038/nature11295. Retrieved 2013-05-29.

- "Flexible strategies for long-term sustainability under uncertainty". Building Research. 40: 545–557. 2012. doi:10.1080/09613218.2012.702565.

- The World Bank, 2006 "Reengaging in Agricultural Water Management: Challenges and Options". pp. 4–5. Retrieved 2011-10-30.

- The World Bank, 2010 "Sustaining water for all in a changing climate: World Bank Group Implementation Progress Report". Retrieved 2011-10-24.

- Chartres, C. and Varma, S. Out of water. From Abundance to Scarcity and How to Solve the World's Water Problems FT Press (USA), 2010.

- Walmsly, N., & Pearce, G. (2010). Towards Sustainable Water Resources Management: Bringing the Strategic Approach up-to-date. Irrigation & Drainage Systems, 24(3/4), 191-203.

Various Water Resources

Ground water is an important water resource. Ground water is present beneath the surface of the Earth and is naturally renewed. This chapter also elucidates seawater and surface water and provides a plethora of interdisciplinary topics for a better comprehension of water resources.

Groundwater

Groundwater (or ground water) is the water present beneath Earth's surface in soil pore spaces and in the fractures of rock formations. A unit of rock or an unconsolidated deposit is called an aquifer when it can yield a usable quantity of water. The depth at which soil pore spaces or fractures and voids in rock become completely saturated with water is called the water table. Groundwater is recharged from, and eventually flows to, the surface naturally; natural discharge often occurs at springs and seeps, and can form oases or wetlands. Groundwater is also often withdrawn for agricultural, municipal, and industrial use by constructing and operating extraction wells. The study of the distribution and movement of groundwater is hydrogeology, also called groundwater hydrology.

The entire surface water flow of the Alapaha River near Jennings, Florida going into a sinkhole leading to the Floridan Aquifer groundwater

Typically, groundwater is thought of as water flowing through shallow aquifers, but, in the technical sense, it can also contain soil moisture, permafrost (frozen soil), immobile water in very low permeability bedrock, and deep geothermal or oil formation water. Groundwater is hypothesized to provide lubrication that can possibly influence the movement of faults. It is likely that much of Earth's subsurface contains some water, which may be mixed with other fluids in some instances. Groundwater may not be confined only to Earth. The formation of some of the landforms observed on Mars may have been influenced by groundwater. There is also evidence that liquid water may also exist in the subsurface of Jupiter's moon Europa.

Groundwater is often cheaper, more convenient and less vulnerable to pollution than surface water. Therefore, it is commonly used for public water supplies. For example, groundwater provides the largest source of usable water storage in the United States, and California annually withdraws the largest amount of groundwater of all the states. Underground reservoirs contain far more water than the capacity of all surface reservoirs and lakes in the US, including the Great Lakes. Many municipal water supplies are derived solely from groundwater.

Polluted groundwater is less visible, but more difficult to clean up, than pollution in rivers and lakes. Groundwater pollution most often results from improper disposal of wastes on land. Major sources include industrial and household chemicals and garbage landfills, excessive fertilizers and pesticides used in agriculture, industrial waste lagoons, tailings and process wastewater from mines, industrial fracking, oil field brine pits, leaking underground oil storage tanks and pipelines, sewage sludge and septic systems.

Aquifers

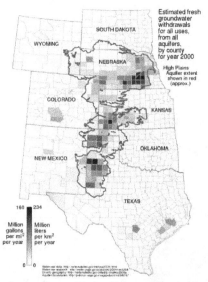

Groundwater withdrawal rates from the Ogallala Aquifer in the Central United States

An *aquifer* is a layer of porous substrate that contains and transmits groundwater. When water can flow directly between the surface and the saturated zone of an aquifer, the aquifer is unconfined. The deeper parts of unconfined aquifers are usually more saturated since gravity causes water to flow downward.

The upper level of this saturated layer of an unconfined aquifer is called the *water table* or *phreatic surface*. Below the water table, where in general all pore spaces are saturated with water, is the phreatic zone.

Substrate with low porosity that permits limited transmission of groundwater is known as an *aquitard*. An *aquiclude* is a substrate with porosity that is so low it is virtually impermeable to groundwater.

A *confined aquifer* is an aquifer that is overlain by a relatively impermeable layer of rock or substrate such as an aquiclude or aquitard. If a confined aquifer follows a downward grade from its *recharge zone*, groundwater can become pressurized as it flows. This can create artesian wells that flow freely without the need of a pump and rise to a higher elevation than the static water table at the above, unconfined, aquifer.

The characteristics of aquifers vary with the geology and structure of the substrate and topography in which they occur. In general, the more productive aquifers occur in sedimentary geologic formations. By comparison, weathered and fractured crystalline rocks yield smaller quantities of groundwater in many environments. Unconsolidated to poorly cemented alluvial materials that have accumulated as valley-filling sediments in major river valleys and geologically subsiding structural basins are included among the most productive sources of groundwater.

The high specific heat capacity of water and the insulating effect of soil and rock can mitigate the effects of climate and maintain groundwater at a relatively steady temperature. In some places where groundwater temperatures are maintained by this effect at about 10 °C (50 °F), groundwater can be used for controlling the temperature inside structures at the surface. For example, during hot weather relatively cool groundwater can be pumped through radiators in a home and then returned to the ground in another well. During cold seasons, because it is relatively warm, the water can be used in the same way as a source of heat for heat pumps that is much more efficient than using air.

The volume of groundwater in an aquifer can be estimated by measuring water levels in local wells and by examining geologic records from well-drilling to determine the extent, depth and thickness of water-bearing sediments and rocks. Before an investment is made in production wells, test wells may be drilled to measure the depths at which water is encountered and collect samples of soils, rock and water for laboratory analyses. Pumping tests can be performed in test wells to determine flow characteristics of the aquifer.

Water Cycle

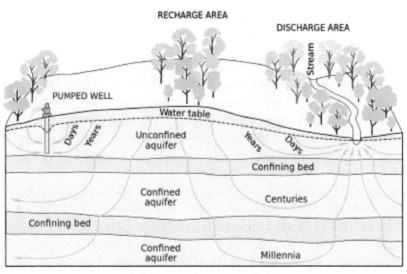

Relative groundwater travel times.

Dzherelo, a common source of drinking water in a Ukrainian village

Groundwater makes up about twenty percent of the world's fresh water supply, which is about 0.61% of the entire world's water, including oceans and permanent ice. Global groundwater storage is roughly equal to the total amount of freshwater stored in the snow and ice pack, including the north and south poles. This makes it an important resource that can act as a natural storage that can buffer against shortages of surface water, as in during times of drought.

Groundwater is naturally replenished by surface water from precipitation, streams, and rivers when this recharge reaches the water table.

Groundwater can be a long-term 'reservoir' of the natural water cycle (with residence times from days to millennia), as opposed to short-term water reservoirs like the atmosphere and fresh surface water (which have residence times from minutes to years). The figure shows how deep groundwater (which is quite distant from the surface recharge) can take a very long time to complete its natural cycle.

The Great Artesian Basin in central and eastern Australia is one of the largest confined aquifer systems in the world, extending for almost 2 million km². By analysing the trace elements in water sourced from deep underground, hydrogeologists have been able to determine that water extracted from these aquifers can be more than 1 million years old.

By comparing the age of groundwater obtained from different parts of the Great Artesian Basin, hydrogeologists have found it increases in age across the basin. Where water recharges the aquifers along the Eastern Divide, ages are young. As groundwater flows westward across the continent, it increases in age, with the oldest groundwater occurring in the western parts. This means that in order to have travelled almost 1000 km from the source of recharge in 1 million years, the groundwater flowing through the Great Artesian Basin travels at an average rate of about 1 metre per year.

Recent research has demonstrated that evaporation of groundwater can play a significant role in the local water cycle, especially in arid regions. Scientists in Saudi Arabia have proposed plans to recapture and recycle this evaporative moisture for crop irrigation. In the opposite photo, a 50-centimeter-square reflective carpet, made of small adjacent plastic cones, was placed in a plant-free dry desert area for five months, without rain or irrigation. It managed to capture and condense enough ground vapor to bring to life naturally buried seeds underneath it, with a green area of about 10% of the carpet area. It is expected that, if seeds were put down before placing this carpet, a much wider area would become green.

Reflective carpet trapping soil water vapor

Issues

Overview

Certain problems have beset the use of groundwater around the world. Just as river waters have been over-used and polluted in many parts of the world, so too have aquifers. The big difference is that aquifers are out of sight. The other major problem is that water management agencies, when calculating the "sustainable yield" of aquifer and river water, have often counted the same water twice, once in the aquifer, and once in its connected river. This problem, although understood for centuries, has persisted, partly through inertia within government agencies. In Australia, for example, prior to the statutory reforms initiated by the Council of Australian Governments water reform framework in the 1990s, many Australian states managed groundwater and surface water through separate government agencies, an approach beset by rivalry and poor communication.

In general, the time lags inherent in the dynamic response of groundwater to development have been ignored by water management agencies, decades after scientific understanding of the issue was consolidated. In brief, the effects of groundwater overdraft (although undeniably real) may take decades or centuries to manifest themselves. In a classic study in 1982, Bredehoeft and colleagues modeled a situation where groundwater extraction in an intermontane basin withdrew the entire annual recharge, leaving 'nothing' for the natural groundwater-dependent vegetation community. Even when the borefield was situated close to the vegetation, 30% of the original vegetation demand could still be met by the lag inherent in the system after 100 years. By year 500, this had reduced to 0%, signalling complete death of the groundwater-dependent vegetation. The science has been available to make these calculations for decades; however, in general water management agencies have ignored effects that will appear outside the rough timeframe of political elections (3 to 5 years). Marios Sophocleous argued strongly that management agencies must define and use appropriate timeframes in groundwater planning. This will mean calculating groundwater withdrawal permits based on predicted effects decades, sometimes centuries in the future.

As water moves through the landscape, it collects soluble salts, mainly sodium chloride. Where such water enters the atmosphere through evapotranspiration, these salts are left behind. In irrigation districts, poor drainage of soils and surface aquifers can result in water tables' coming to the surface in low-lying areas. Major land degradation problems of soil salinity and waterlogging result, combined with increasing levels of salt in surface waters. As a consequence, major damage has occurred to local economies and environments.

Four important effects are worthy of brief mention. First, flood mitigation schemes, intended to protect infrastructure built on floodplains, have had the unintended consequence of reducing aquifer recharge associated with natural flooding. Second, prolonged depletion of groundwater in extensive aquifers can result in land subsidence, with associated infrastructure damage – as well as, third, saline intrusion. Fourth, draining acid sulphate soils, often found in low-lying coastal plains, can result in acidification and pollution of formerly freshwater and estuarine streams.

Another cause for concern is that groundwater drawdown from over-allocated aquifers has the potential to cause severe damage to both terrestrial and aquatic ecosystems – in some cases very conspicuously but in others quite imperceptibly because of the extended period over which the damage occurs.

Overdraft

Wetlands contrast the arid landscape around Middle Spring, Fish Springs National Wildlife Refuge, Utah

Groundwater is a highly useful and often abundant resource. However, over-use, or **overdraft**, can cause major problems to human users and to the environment. The most evident problem (as far as human groundwater use is concerned) is a lowering of the water table beyond the reach of existing wells. As a consequence, wells must be drilled deeper to reach the groundwater; in some places (e.g., California, Texas, and India) the water table has dropped hundreds of feet because of extensive well pumping. In the Punjab region of India, for example, groundwater levels have dropped 10 meters since 1979, and the rate of depletion is accelerating. A lowered water table may, in turn, cause other problems such as groundwater-related subsidence and saltwater intrusion.

Groundwater is also ecologically important. The importance of groundwater to ecosystems is often overlooked, even by freshwater biologists and ecologists. Groundwaters sustain rivers, wetlands, and lakes, as well as subterranean ecosystems within karst or alluvial aquifers.

Not all ecosystems need groundwater, of course. Some terrestrial ecosystems – for example, those of the open deserts and similar arid environments – exist on irregular rainfall and the moisture it delivers to the soil, supplemented by moisture in the air. While there are other terrestrial ecosystems in more hospitable environments where groundwater plays no central role, groundwater is in fact fundamental to many of the world's major ecosystems. Water flows between groundwaters and surface waters. Most rivers, lakes, and wetlands are fed by, and (at other places or times) feed groundwater, to varying degrees. Groundwater feeds soil moisture through percolation, and many terrestrial vegetation communities depend directly on either groundwater or the percolated soil moisture above the aquifer for at least part of each year. Hyporheic zones (the mixing zone of streamwater and groundwater) and riparian zones are examples of ecotones largely or totally dependent on groundwater.

Subsidence

Subsidence occurs when too much water is pumped out from underground, deflating the space below the above-surface, and thus causing the ground to collapse. The result can look like craters on plots of land. This occurs because, in its natural equilibrium state, the hydraulic pressure of groundwater in the pore spaces of the aquifer and the aquitard supports some of the weight of the overlying sediments. When groundwater is removed from aquifers by excessive pumping, pore pressures in the aquifer drop and compression of the aquifer may occur. This compression may be partially recoverable if pressures rebound, but much of it is not. When the aquifer gets compressed, it may cause land subsidence, a drop in the ground surface. The city of New Orleans, Louisiana is actually below sea level today, and its subsidence is partly caused by removal of groundwater from the various aquifer/aquitard systems beneath it. In the first half of the 20th century, the San Joaquin Valley experienced significant subsidence, in some places up to 8.5 metres (28 feet) due to groundwater removal. Cities on river deltas, including Venice in Italy, and Bangkok in Thailand, have experienced surface subsidence; Mexico City, built on a former lake bed, has experienced rates of subsidence of up to 40 cm (1'3") per year.

Seawater Intrusion

In general, in very humid or undeveloped regions, the shape of the water table mimics the slope of the surface. The recharge zone of an aquifer near the seacoast is likely to be inland, often at considerable distance. In these coastal areas, a lowered water table may induce sea water to reverse the flow toward the land. Sea water moving inland is called a saltwater intrusion. In alternative fashion, salt from mineral beds may leach into the groundwater of its own accord.

Pollution

Polluted groundwater is less visible, but more difficult to clean up, than pollution in rivers and lakes. Groundwater pollution most often results from improper disposal of wastes on land. Major sources include industrial and household chemicals and garbage landfills, industrial waste lagoons, tailings and process wastewater from mines, oil field brine pits, leaking underground oil storage tanks and pipelines, sewage sludge and septic systems. Polluted groundwater is mapped by sampling soils and groundwater near suspected or known sources of pollution, to determine the extent of the pollution, and to aid in the design of groundwater remediation systems. Preventing groundwater pollution near potential sources such as landfills requires lining the bottom of a landfill with watertight materials, collecting any leachate with drains, and keeping rainwater off any potential contaminants, along with regular monitoring of nearby groundwater to verify that contaminants have not leaked into the groundwater.

Iron oxide staining caused by reticulation from an unconfined aquifer in karst topography. Perth, Western Australia.

Groundwater pollution, from pollutants released to the ground that can work their way down into groundwater, can create a contaminant plume within an aquifer. Pollution can occur from landfills, naturally occurring arsenic, on-site sanitation systems or other point sources, such as petrol stations or leaking sewers.

Movement of water and dispersion within the aquifer spreads the pollutant over a wider area, its advancing boundary often called a plume edge, which can then intersect with groundwater wells or daylight into surface water such as seeps and springs, making the water supplies unsafe for humans and wildlife. Different mechanism have influence on the transport of pollutants, e.g. diffusion, adsorption, precipitation, decay, in the groundwater. The interaction of groundwater contamination with surface waters is analyzed by use of hydrology transport models.

The danger of pollution of municipal supplies is minimized by locating wells in areas of deep groundwater and impermeable soils, and careful testing and monitoring of the aquifer and nearby potential pollution sources.

Government Regulations

In the United States, laws regarding ownership and use of groundwater are generally state laws; however, regulation of groundwater to minimize pollution of groundwater is by both states and the federal-level Environmental Protection Agency. Ownership and use rights to groundwater typically follow one of three main systems:

Rule of Capture

The Rule of Capture provides each landowner the ability to capture as much groundwater as they can put to a beneficial use, but they are not guaranteed any set amount of water. As a result, well-owners are not liable to other landowners for taking water from beneath their land. State laws or regulations will often define "beneficial use", and sometimes place other limits, such as disallowing groundwater extraction which causes subsidence on neighboring property.

Riparian Rights

Limited private ownership rights similar to riparian rights in a surface stream. The amount of groundwater right is based on the size of the surface area where each landowner gets a corresponding amount of the available water. Once adjudicated, the maximum amount of the water right is set, but the right can be decreased if the total amount of available water decreases as is likely during a drought. Landowners may sue others for encroaching upon their groundwater rights, and water pumped for use on the overlying land takes preference over water pumped for use off the land.

Environmental Protection of Groundwater

In November 2006, the Environmental Protection Agency published the groundwater Rule in the United States Federal Register. The EPA was worried that the groundwater system would be vulnerable to contamination from fecal matter. The point of the rule was to keep microbial pathogens out of public water sources. The 2006 groundwater Rule was an amendment of the 1996 Safe Drinking Water Act.

Others

Reasonable Use Rule (American Rule)

This rule does not guarantee the landowner a set amount of water, but allows unlimited extraction as long as the result does not unreasonably damage other wells or the aquifer system. Usually this rule gives great weight to historical uses and prevents new uses that interfere with the prior use.

Groundwater Scrutiny Upon Real Estate Property Transactions in the Us

In the US, upon commercial real estate property transactions both groundwater and soil are the subjects of scrutiny, with a Phase I Environmental Site Assessment normally being prepared to investigate and disclose potential pollution issues. In the San Fernando Valley of California, real estate contracts for property transfer below the Santa Susana Field Laboratory (SSFL) and eastward have clauses releasing the seller from liability for groundwater contamination consequences from existing or future pollution of the Valley Aquifer.

Sub-field of Ground Water

Endorheic Basin

An endorheic basin is a closed drainage basin that retains water and allows no outflow to other external bodies of water, such as rivers or oceans, but converges instead into lakes or swamps, permanent or seasonal, that equilibrate through evaporation. Such a basin may also be referred to as a closed or terminal basin or as an internal drainage system.

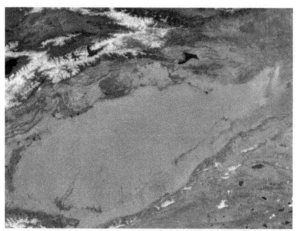

NASA photo of the endorheic Tarim Basin

Endorheic basin showing waterflow input into Üüreg Lake

Normally, water that has accrued in a drainage basin eventually flows out through rivers or streams on the Earth's surface or by underground diffusion through permeable rock, ultimately ending up in the oceans. However, in an endorheic basin, rain (or other precipitation) that falls within it does not flow out but may only leave the drainage system by evaporation and seepage. The bottom of such a basin is typically occupied by a salt lake or salt pan.

Endorheic regions, in contrast to exorheic regions which flow to the ocean in geologically defined patterns, are closed hydrologic systems. Their surface waters drain to inland terminal locations where the water evaporates or seeps into the ground, having no access to discharge into the sea. Endorheic water bodies include some of the largest lakes in the world, such as the Aral Sea (formerly) and the Caspian Sea, the world's largest saline inland sea.

Most endorheic basins are arid, although there are many notable exceptions, such as the Valley of Mexico, the Lake Tahoe region, and various regions in the Caspian Basin.

Endorheic basins can be massively and rapidly affected by climate change and excessive water abstraction. An exorheic lake naturally remains at an overflow level, so water flow into the lake may be many times more than is needed to maintain its present size. In contrast, an endorheic basin does not have enough inflow that it overflows to the ocean, so any loss of water intake can immediately begin to shrink the lake. In the past century or so, many very large endorheic lakes have been reduced to small remnants of their former size, as with Lake Chad and Lake Urmia, or gone completely as have Tulare Lake and Fucine Lake. The same effect was seen at the end of the Ice Age, in which many extremely large lakes in the Sahara and western United States disappeared or were drastically reduced, leaving behind desert basins, salt pans and remnant saline lakes.

Endorheic Lakes

Endorheic lakes are bodies of water that do not flow into the sea. Most of the water falling on Earth finds its way to the oceans through a network of rivers, lakes and wetlands. However, there is a class of water bodies that are located in closed or endorheic watersheds where the topography prevents their drainage to the oceans. These endorheic watersheds (containing water in rivers or lakes that form a balance of surface inflows, evaporation and seepage) are often called terminal lakes or sink lakes.

Endorheic lakes are usually in the interior of a landmass, far from an ocean in areas of relatively low rainfall. Their watersheds are often confined by natural geologic land formations such as a mountain range, cutting off water egress to the ocean. The inland water flows into dry watersheds where the water evaporates, leaving a high concentration of minerals and other inflow erosion products. Over time this input of erosion products can cause the endorheic lake to become relatively saline (a "salt lake"). Since the main outflow pathways of these lakes are chiefly through evaporation and seepage, endorheic lakes are usually more sensitive to environmental pollutant inputs than water bodies that have access to oceans, as pollution can be trapped in them and accumulate over time.

Occurrence

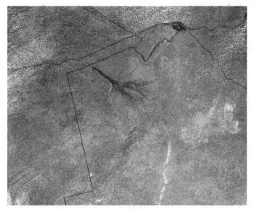

The Okavango Delta (centre) of southern Africa, where the Okavango River spills out into the empty trough of the Kalahari Desert. The area was a lake fed by the river during the Ice Ages.

Endorheic regions can occur in any climate but are most commonly found in desert locations. In areas where rainfall is higher, riparian erosion will generally carve drainage channels (particularly in times of flood), or cause the water level in the terminal lake to rise until it finds an outlet, breaking the enclosed endorheic hydrological system's geographical barrier and opening it to the surrounding terrain. The Black Sea was likely such a lake, having once been an independent hydrological system before the Mediterranean Sea broke through the terrain separating the two. Lake Bonneville was another such lake, overflowing its basin in the Bonneville flood. Malheur/Harney Lake in Oregon is normally cut off from drainage to the ocean, but has an outflow channel to the Malheur River that is normally dry, but flows in years of peak precipitation.

Examples of relatively humid regions in endorheic basins often exist at high elevation. These regions tend to be marshy and are subject to substantial flooding in wet years. The area containing Mexico City is one such case, with annual precipitation of 850 mm and characterized by water-logged soils that require draining.

Endorheic regions tend to be far inland with their boundaries defined by mountains or other geological features that block their access to oceans. Since the inflowing water can evacuate only through seepage or evaporation, dried minerals or other products collect in the basin, eventually making the water saline and also making the basin vulnerable to pollution. Continents vary in their concentration of endorheic regions due to conditions of geography and climate. Australia has the highest percentage of endorheic regions at 21 percent while North America has the least at 5 percent. Approximately 18 percent of the earth's land drains to endorheic lakes or seas, the largest of these land areas being the interior of Asia.

In deserts, water inflow is low and loss to solar evaporation high, drastically reducing the formation of complete drainage systems. Closed water flow areas often lead to the concentration of salts and other minerals in the basin. Minerals leached from the surrounding rocks are deposited in the basin, and left behind when the water evaporates. Thus endorheic basins often contain extensive salt pans (also called salt flats, salt lakes, alkali flats, dry lake beds or playas). These areas tend to be large, flat hardened surfaces and are sometimes used for aviation runways or land speed record attempts, because of their extensive areas of perfectly level terrain.

Both permanent and seasonal endorheic lakes can form in endorheic basins. Some endorheic basins are essentially stable, climate change having reduced precipitation to the degree that a lake no longer forms. Even most permanent endorheic lakes change size and shape dramatically over time, often becoming much smaller or breaking into several smaller parts during the dry season. As humans have expanded into previously uninhabitable desert areas, the river systems that feed many endorheic lakes have been altered by the construction of dams and aqueducts. As a result, many endorheic lakes in developed or developing countries have contracted dramatically, resulting in increased salinity, higher concentrations of pollutants, and the disruption of ecosystems.

Even within exorheic basins, there can be "non-contributing", low-lying areas that trap runoff and prevent it from contributing to flows downstream during years of average or below-average runoff. In flat river basins, non-contributing areas can be a large fraction of the river basin, e.g. Lake Winnipeg's basin. A lake may be endorheic during dry years and can overflow its basin during wet years, e.g., the former Tulare Lake.

Because the Earth's climate has recently been through a warming and drying phase with the end of the Ice Ages, many endorheic areas such as Death Valley that are now dry deserts were large lakes relatively recently. During the last ice age, the Sahara may have contained lakes larger than any now existing.

Notable Endorheic Basins and Lakes

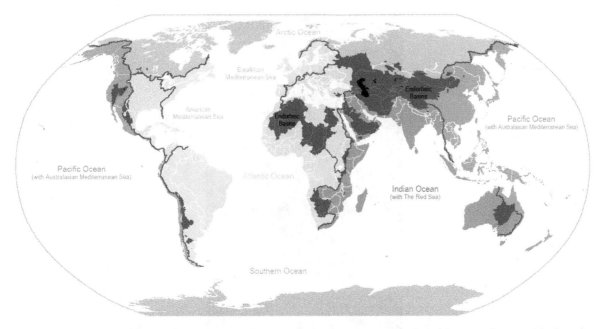

Major endorheic basins of the world. Basins are shown in dark gray; major endorheic lakes are shown in black. Colored regions represent the major drainage patterns of the continents to the oceans (non-endorheic). Continental divides are indicated by dark lines.

Antarctica

Endorheic lakes in Antarctica are located in the McMurdo Dry Valleys, Victoria Land, the largest ice-free area in Antarctica.

- Don Juan Pond in Wright Valley is fed by groundwater from a rock glacier and remains unfrozen throughout the year.

- Lake Vanda in Wright Valley has a perennial ice cover, the edges of which melt in the summer allowing flow from the longest river in Antarctica, the Onyx River. The lake is over 70 m deep and is hypersaline.

- Lake Bonney is in Taylor Valley and has a perennial ice cover and two lobes separated by the Bonney Riegel. The lake is fed by glacial melt and discharge from Blood Falls. Its unique glacial history has resulted in a hypersaline brine in the bottom waters and fresh water at the surface.

- Lake Hoare, in Taylor Valley, is the freshest of the Dry Valley lakes receiving its melt almost exclusively from the Canada Glacier. The lake has an ice cover and forms a moat during the Austral summer.

- Lake Fryxell is adjacent to the Ross Sea in Taylor Valley. The lake has an ice cover and receives its water from numerous glacial meltwater streams for approximately 6 weeks out of the year. Its salinity increases with depth.

Asia

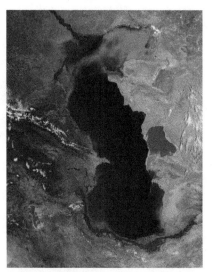

Caspian Sea, a giant inland basin

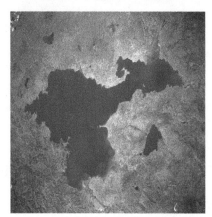

Lake Van, Turkey

Much of western and Central Asia is a giant endorheic region made up of a number of contiguous closed basins. The region contains several basins and terminal lakes, including:

- The Caspian Sea, the largest lake on Earth. A large part of Eastern Europe, drained by the Volga River, is part of the Caspian's basin.

- Lake Urmia in Western Azerbaijan Province of Iran.

- The Aral Sea, whose tributary rivers have been diverted, leading to a dramatic shrinkage of the lake. The resulting ecological disaster has brought the plight faced by internal drainage basins to public attention.

- Lake Balkhash, in Kazakhstan.

- Issyk-Kul Lake, Son-Kul Lake, and Chatyr-Kul Lake in Kyrgyzstan.

- Lop Lake, in the Tarim Basin of China's Xinjiang Uygur Autonomous Region.

- The Dzungarian Basin in Xinjiang, separated from the Tarim Basin by the Tian Shan. The most notable terminal lake in the basin is the Manas Lake.

- The Central Asian Internal Drainage Basin, in southern and western Mongolia, contains a series of closed drainage basins, such as the Khyargas Nuur basin, the Uvs Nuur basin, and the Pu-Lun-To River Basin.

- Qaidam Basin, in Qinghai Province, China, as well as nearby Qinghai Lake.

- Sistan Basin covering areas of Iran and Afghanistan

- Pangong Tso on the China-India border

- Many small lakes and rivers of the Iranian Plateau, including Gavkhouni marshes and Namak Lake.

Other endorheic lakes and basins in Asia include:

- The Dead Sea, the lowest surface point on Earth and one of its saltiest bodies of water, lies between Israel and Jordan.

- Sambhar Lake, in Rajasthan, north-western India

- Lake Van in eastern Turkey

- Sabkhat al-Jabbul, extensive salt flats and a 100 square kilometres (39 sq mi) lake in Syria.

- Solar Lake, Sinai, near the Israeli-Egypt border.

- Lake Tuz, in Turkey, in south part of Central Anatolia Region.

- Sawa lake in Iraq, in Muthanna Governorate.

Australia

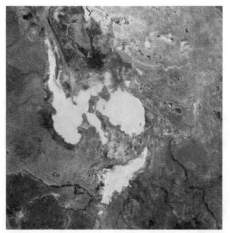

A false-colour satellite photo of Australia's Lake Eyre Image credit: *NASA's Earth Observatory*

Australia, being very dry and having exceedingly low runoff ratios due to its ancient soils, has many endorheic drainages. The most important are:

- Lake Eyre Basin, which drains into the highly variable Lake Eyre and includes Lake Frome.
- Lake Torrens, to the west of the Flinders Ranges in South Australia.
- Lake Corangamite, a highly saline crater lake in western Victoria.
- Lake George, formerly connected to the Murray-Darling Basin

Africa

Large endorheic regions in Africa are located in the Sahara Desert, the Kalahari Desert, and the East African Rift:

- Chad Basin, in the northern center of Africa. It covers an area of approximately 2.434 million km^2.
- Qattara Depression, in Egypt.
- Chott Melrhir, in Algeria.
- Chott el Djerid, in Tunisia.
- The Okavango River, in the Kalahari Desert, is part of an endorheic basin region which also includes the Okavango Delta, Lake Ngami, the Nata River, and a number of salt pans such as Makgadikgadi Pan.
- Etosha pan in Namibia's Etosha National Park.
- Lake Turkana, in Kenya, whose basin includes the Omo River of Ethiopia.
- Lake Chilwa, in Malawi.
- Afar Depression, in Eritrea, Ethiopia, and Djibouti, which contains the Awash River
- Some Rift Valley lakes, such as Lake Abijatta, Lake Chew Bahir, Lake Shala, Lake Chamo, and Lake Awasa.
- Lake Mweru Wantipa, in Zambia.
- Lake Magadi, in Kenya.

North and Central America

The dry lake in the Badwater Basin in Death Valley National Park.

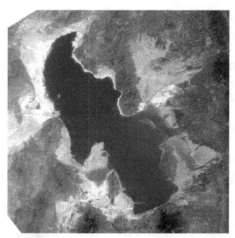

Great Salt Lake, Satellite photo (2003) after five years of drought

- The Valley of Mexico. In Pre-Columbian times, the Valley was substantially covered with five lakes, including Lake Texcoco, Lake Xochimilco, and Lake Chalco.

- Guzmán Basin, in northern Mexico and the southwestern United States. The Mimbres River of New Mexico drains into this basin.

- Lago de Atitlán, in the highlands of Guatemala.

- Lago de Coatepeque, El Salvador.

- Bolsón de Mapimí, in northern Mexico.

- Willcox Playa of southern Arizona.

- The Great Basin, which covers much of Nevada, Oregon and Utah, is a large region of contiguous closed basins, including:

  o Great Salt Lake, in Utah, the largest terminal lake in the Western Hemisphere.

  o The Black Rock Desert in Nevada, location of the Thrust2 and ThrustSSC landspeed record runs, and the annual home to the Burning Man festival.

  o Death Valley, the lowest land point in the United States. During the Holocene epoch, Lake Manly filled the Death Valley basin of Inyo County, California. Later, Death Valley had a system of tributary basins:

    ▪ Rogers Dry Lake, at Edwards Air Force Base in California.

    ▪ Owens Lake and the Owens River basin.

      ▪ Mono Lake in California.

  o Groom Dry Lake in Nevada, location of Area 51.

  o Abert Lake and Summer Lake in Oregon.

  o Goose Lake (Oregon-California) on the California/Oregon border. Historically it drained into the Pit River. Agricultural development and irrigation diversions have lowered the lake level so that it no longer drains to the sea.

- o Malheur Lake in Oregon.

- o Warner Valley in Oregon.

- o Alvord Desert in Oregon.

- o Salton Sea in California, a lake accidentally recreated in 1905 when irrigation canals ruptured, filling a desert endorheic basin and recreating an ancient saline sea.

- o Lake Elsinore in California and the San Jacinto River. Intermittently drains into the Santa Ana River.

- o Sevier Lake, in Utah.

- o Pyramid Lake in Nevada, whose drainage basin includes Lake Tahoe.

- o Humboldt Sink and the Humboldt River basin in Nevada.

- o Carson Sink and the Carson River basin in Nevada.

- o Walker Lake and the Walker River in Nevada.

- Tulare Lake in the southern end of the San Joaquin Valley fed by the Kaweah and Tule Rivers plus southern distributaries of the Kings. Historically, in very wet years it would drain into the San Joaquin River. Agricultural development and irrigation diversions have left the lake dry.

- Buena Vista Lake at the southmost end of the San Joaquin Valley fed by the Kern River. Historically, in exceptionally wet years it would drain into Tulare Lake and then into the San Joaquin River. Agricultural development and irrigation diversions have left the lake dry.

- Crater Lake, in Oregon. It is filled directly by rain and snow, and so has very little mineral or salt buildup.

- The Great Divide Basin in Wyoming, a small endorheic basin that straddles the Continental Divide of the Americas.

- Devils Lake, in North Dakota.

- Devil's Lake, in Wisconsin.

- Tule Lake and the Lost River basin in California and Oregon.

- Little Manitou Lake in Saskatchewan.

- Old Wives Lake, on the Laurentian Divide in Saskatchewan.

- Pakowki Lake, on the Laurentian Divide in Alberta.

- Spotted Lake, Osoyoos, British Columbia, Canada.

- Frame Lake in Yellowknife, capital of Canada's Northwest Territories.

- New Mexico has a number of desert endorheic basins including:

    o The Tularosa Basin, a rift valley.

    o Zuni Salt Lake, a maar.

    o The Mimbres River Basin, in Grant County.

- Lago Enriquillo on the island of Hispaniola in the Caribbean Sea.

Many small lakes and ponds in North Dakota and the Northern Great Plains are endorheic; some of them have salt encrustations along their shores.

Europe

The Lasithi Plateau in Crete

Though a large portion of Europe (about 19%, located in Russia and Kazakhstan) drains to the endorheic Caspian Sea, Europe's wet climate means it contains relatively few terminal lakes itself: any such basin is likely to continue to fill until it reaches an overflow level connecting it with an outlet or erodes the barrier blocking its exit. Exceptions include:

- Lake Neusiedl, in Austria and Hungary.

- Lake Trasimeno, in Italy.

- Fucine Lake, in Italy. Now drained.

- Lake Velence, in Hungary.

- Lake Prespa, between Albania, Greece and the Republic of Macedonia.

- Rahasane turlough, the largest turlough in Ireland.

- Laacher, in Germany.

- The Lasithi Plateau in Crete, Greece, is a high endorheic plateau.

All these lakes are drained, however, either through manmade canals or via karstic phenomena. Minor endorheic lakes exist throughout the Mediterranean countries of Spain (e.g. Laguna de Gallocanta, Estany de Banyoles), Italy, Cyprus (Larnaca and Akrotiri salt lakes) and Greece.

South America

MODIS image from November 4, 2001 showing Lake Titicaca, the Salar de Uyuni, and the Salar de Coipasa. These are all parts of the Altiplano

- Laguna del Carbón, in Gran Bajo de San Julián, Argentina – the lowest point in the Western and Southern hemispheres

- Lake Mar Chiquita in Argentina.

- The Altiplano includes a number of closed basins such as Poopó Lake.

- Lake Valencia, in Venezuela.

- Salar de Atacama, in the Atacama Desert, Chile.

Ancient

Some of the Earth's ancient endorheic systems and lakes include:

- The Black Sea, until its merger with the Mediterranean.

- The Mediterranean Sea itself and all its tributary basins, during its Messinian desiccation (approximately five million years ago) as it became disconnected from the Atlantic Ocean.

- The Orcadian Basin in Scotland during the Devonian period. Now identifiable as lacustrine sediments buried around and off the coast.

- Lake Tanganyika in Africa. Currently high enough to connect to rivers entering the sea.

- Lake Lahontan in North America.

- Lake Bonneville in North America. Basin was not always endorheic; at times overflowed through Red Rock Pass to the Snake River and the sea.

- Lake Chewaucan in North America.

- Tularosa Basin and Lake Cabeza de Vaca in North America. Basin formerly much larger than at present, including the ancestral Rio Grande north of Texas, feeding a large lake area.

- Ebro and Duero basins, draining most of northern Spain during the Neogene and perhaps Pliocene. Climate change and erosion of the Catalan coastal mountains, as well as deposition of alluvium in the terminal lake, allowed the Ebro basin to overflow into the sea during the middle-to-late Miocene.

Irrigation

An irrigation sprinkler watering a lawn

Irrigation canal in Osmaniye, Turkey

Irrigation is the method in which water is supplied to plants at regular intervals for agriculture. It is used to assist in the growing of agricultural crops, maintenance of landscapes, and revegetation of disturbed soils in dry areas and during periods of inadequate rainfall. Additionally, irrigation also has a few other uses in crop production, which include protecting plants against frost, suppressing weed growth in grain fields and preventing soil consolidation. In contrast, agriculture that relies only on direct rainfall is referred to as rain-fed or dry land farming.

Irrigation systems are also used for dust suppression, disposal of sewage, and in mining. Irrigation is often studied together with drainage, which is the natural or artificial removal of surface and sub-surface water from a given area.

Irrigation has been a central feature of agriculture for over 5,000 years and is the product of many cultures. Historically, it was the basis for economies and societies across the globe, from Asia to the Southwestern United States.

History

Animal-powered irrigation, Upper Egypt, ca. 1846

An example of an irrigation system common on the Indian subcontinent. Artistic impression on the banks of Dal Lake, Kashmir, India

Inside a karez tunnel at Turpan, Sinkiang

Archaeological investigation has identified as evidence of irrigation where the natural rainfall was insufficient to support crops for rainfed agriculture.

Perennial irrigation was practiced in the Mesopotamian plain whereby crops were regularly watered throughout the growing season by coaxing water through a matrix of small channels formed in the field.

Irrigation in Tamil Nadu (India)

Ancient Egyptians practiced *Basin irrigation* using the flooding of the Nile to inundate land plots which had been surrounded by dykes. The flood water was held until the fertile sediment had settled before the surplus was returned to the watercourse. There is evidence of the ancient Egyptian pharaoh Amenemhet III in the twelfth dynasty (about 1800 BCE) using the natural lake of the Faiyum Oasis as a reservoir to store surpluses of water for use during the dry seasons, the lake swelled annually from flooding of the Nile.

The Ancient Nubians developed a form of irrigation by using a waterwheel-like device called a *sakia*. Irrigation began in Nubia some time between the third and second millennium BCE. It largely depended upon the flood waters that would flow through the Nile River and other rivers in what is now the Sudan.

In sub-Saharan Africa irrigation reached the Niger River region cultures and civilizations by the first or second millennium BCE and was based on wet season flooding and water harvesting.

Terrace irrigation is evidenced in pre-Columbian America, early Syria, India, and China. In the Zana Valley of the Andes Mountains in Peru, archaeologists found remains of three irrigation canals radiocarbon dated from the 4th millennium BCE, the 3rd millennium BCE and the 9th century CE. These canals are the earliest record of irrigation in the New World. Traces of a canal possibly dating from the 5th millennium BCE were found under the 4th millennium canal. Sophisticated irrigation and storage systems were developed by the Indus Valley Civilization in present-day Pakistan and North India, including the reservoirs at Girnar in 3000 BCE and an early canal irrigation system from circa 2600 BCE. Large scale agriculture was practiced and an extensive network of canals was used for the purpose of irrigation.

Ancient Persia (modern day Iran) as far back as the 6th millennium BCE, where barley was grown in areas where the natural rainfall was insufficient to support such a crop. The Qanats, developed in ancient Persia in about 800 BCE, are among the oldest known irrigation methods still in use today. They are now found in Asia, the Middle East and North Africa. The system comprises a network of vertical wells and gently sloping tunnels driven into the sides of cliffs and steep hills to tap groundwater. The noria, a water wheel with clay pots around the rim powered by the flow of the stream (or by animals where the water source was still), was first brought into use at about this time, by Roman settlers in North Africa. By 150 BCE the pots were fitted with valves to allow smoother filling as they were forced into the water.

The irrigation works of ancient Sri Lanka, the earliest dating from about 300 BCE, in the reign of King Pandukabhaya and under continuous development for the next thousand years, were one of the most complex irrigation systems of the ancient world. In addition to underground canals, the Sinhalese were the first to build completely artificial reservoirs to store water. Due to their engineering superiority in this sector, they were often called 'masters of irrigation'. Most of these irrigation systems still exist undamaged up to now, in Anuradhapura and Polonnaruwa, because of the advanced and precise engineering. The system was extensively restored and further extended during the reign of King Parakrama Bahu (1153–1186 CE).

China

The oldest known hydraulic engineers of China were Sunshu Ao (6th century BCE) of the Spring and Autumn Period and Ximen Bao (5th century BCE) of the Warring States period, both of whom worked on large irrigation projects. In the Sichuan region belonging to the State of Qin of ancient China, the Dujiangyan Irrigation System was built in 256 BCE to irrigate an enormous area of farmland that today still supplies water. By the 2nd century AD, during the Han Dynasty, the Chinese also used chain pumps that lifted water from lower elevation to higher elevation. These were powered by manual foot pedal, hydraulic waterwheels, or rotating mechanical wheels pulled by oxen. The water was used for public works of providing water for urban residential quarters and palace gardens, but mostly for irrigation of farmland canals and channels in the fields.

Korea

In 15th century Korea, the world's first rain gauge, *uryanggye* (Korean: 우량계), was invented in 1441. The inventor was Jang Yeong-sil, a Korean engineer of the Joseon Dynasty, under the active direction of the king, Sejong the Great. It was installed in irrigation tanks as part of a nationwide system to measure and collect rainfall for agricultural applications. With this instrument, planners and farmers could make better use of the information gathered in the survey.

North America

In North America, the Hohokam were the only culture to rely on irrigation canals to water their crops, and their irrigation systems supported the largest population in the Southwest by AD 1300. The Hohokam constructed an assortment of simple canals combined with weirs in their various agricultural pursuits. Between the 7th and 14th centuries, they also built and maintained extensive irrigation networks along the lower Salt and middle Gila rivers that rivaled the complexity of those used in the ancient Near East, Egypt, and China. These were constructed using relatively simple excavation tools, without the benefit of advanced engineering technologies, and achieved drops of a few feet per mile, balancing erosion and siltation. The Hohokam cultivated varieties of cotton, tobacco, maize, beans and squash, as well as harvested an assortment of wild plants. Late in the Hohokam Chronological Sequence, they also used extensive dry-farming systems, primarily to grow agave for food and fiber. Their reliance on agricultural strategies based on canal irrigation, vital in their less than hospitable desert environment and arid climate, provided the basis for the aggregation of rural populations into stable urban centers.

Present Extent

Irrigation ditch in Montour County, Pennsylvania, off Strawberry Ridge Road

In the mid-20th century, the advent of diesel and electric motors led to systems that could pump groundwater out of major aquifers faster than drainage basins could refill them. This can lead to permanent loss of aquifer capacity, decreased water quality, ground subsidence, and other problems. The future of food production in such areas as the North China Plain, the Punjab, and the Great Plains of the US is threatened by this phenomenon.

At the global scale, 2,788,000 km² (689 million acres) of fertile land was equipped with irrigation infrastructure around the year 2000. About 68% of the area equipped for irrigation is located in Asia, 17% in the Americas, 9% in Europe, 5% in Africa and 1% in Oceania. The largest contiguous areas of high irrigation density are found:

- In Northern India and Pakistan along the Ganges and Indus rivers

- In the Hai He, Huang He and Yangtze basins in China

- Along the Nile river in Egypt and Sudan

- In the Mississippi-Missouri river basin and in parts of California

Smaller irrigation areas are spread across almost all populated parts of the world.

Only eight years later in 2008, the scale of irrigated land increased to an estimated total of 3,245,566 km² (802 million acres), which is nearly the size of India.

Types of Irrigation

Basin flood irrigation of wheat

Irrigation of land in Punjab, Pakistan

Various types of irrigation techniques differ in how the water obtained from the source is distributed within the field. In general, the goal is to supply the entire field uniformly with water, so that each plant has the amount of water it needs, neither too much nor too little.

Surface Irrigation

In *surface* (*furrow, flood*, or *level basin*) irrigation systems, water moves across the surface of agricultural lands, in order to wet it and infiltrate into the soil. Surface irrigation can be subdivided into furrow, *borderstrip or basin irrigation*. It is often called *flood irrigation* when the irrigation results in flooding or near flooding of the cultivated land. Historically, this has been the most common method of irrigating agricultural land and still is in most parts of the world.

Where water levels from the irrigation source permit, the levels are controlled by dikes, usually plugged by soil. This is often seen in terraced rice fields (rice paddies), where the method is used to flood or control the level of water in each distinct field. In some cases, the water is pumped, or lifted by human or animal power to the level of the land. The field water efficiency of surface irrigation is typically lower than other forms of irrigation but has the potential for efficiencies in the range of 70% - 90% under appropriate management.

Localized Irrigation

Impact sprinkler head

Localized irrigation is a system where water is distributed under low pressure through a piped network, in a pre-determined pattern, and applied as a small discharge to each plant or adjacent to it. Drip irrigation, spray or micro-sprinkler irrigation and bubbler irrigation belong to this category of irrigation methods.

Subsurface Textile Irrigation

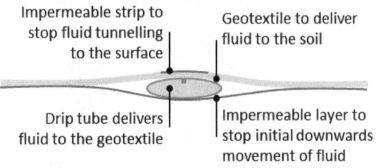

Diagram showing the structure of an example SSTI installation

Subsurface Textile Irrigation (SSTI) is a technology designed specifically for subsurface irrigation in all soil textures from desert sands to heavy clays. A typical subsurface textile irrigation system has an impermeable base layer (usually polyethylene or polypropylene), a drip line running along that base, a layer of geotextile on top of the drip line and, finally, a narrow impermeable layer on top of the geotextile. Unlike standard drip irrigation, the spacing of emitters in the drip pipe is not critical as the geotextile moves the water along the fabric up to 2 m from the dripper.

Drip Irrigation

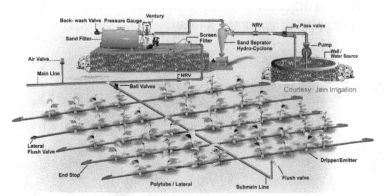

Drip irrigation layout and its parts

Drip irrigation – a dripper in action

Grapes in Petrolina, only made possible in this semi arid area by drip irrigation

Drip (or micro) irrigation, also known as trickle irrigation, functions as its name suggests. In this system water falls drop by drop just at the position of roots. Water is delivered at or near the root zone of plants, drop by drop. This method can be the most water-efficient method of irrigation, if managed properly, since evaporation and runoff are minimized. The field water efficiency of drip irrigation is typically in the range of 80 to 90 percent when managed correctly.

In modern agriculture, drip irrigation is often combined with plastic mulch, further reducing evaporation, and is also the means of delivery of fertilizer. The process is known as *fertigation*.

Deep percolation, where water moves below the root zone, can occur if a drip system is operated for too long or if the delivery rate is too high. Drip irrigation methods range from very high-tech and computerized to low-tech and labor-intensive. Lower water pressures are usually needed than for most other types of systems, with the exception of low energy center pivot systems and surface irrigation systems, and the system can be designed for uniformity throughout a field or for precise water delivery to individual plants in a landscape containing a mix of plant species. Although it is difficult to regulate pressure on steep slopes, pressure compensating emitters are available, so the field does not have to be level. High-tech solutions involve precisely calibrated emitters located along lines of tubing that extend from a computerized set of valves.

Irrigation using Sprinkler Systems

Sprinkler irrigation of blueberries in Plainville, New York, United States

A traveling sprinkler at Millets Farm Centre, Oxfordshire, United Kingdom

In *sprinkler* or overhead irrigation, water is piped to one or more central locations within the field and distributed by overhead high-pressure sprinklers or guns. A system utilizing sprinklers, sprays, or guns mounted overhead on permanently installed risers is often referred to as a *solid-set* irrigation system. Higher pressure sprinklers that rotate are called *rotors* an are driven by a ball drive, gear drive, or impact mechanism. Rotors can be designed to rotate in a full or partial circle. Guns are similar to rotors, except that they generally operate at very high pressures of 40 to 130 lbf/in² (275 to 900 kPa) and flows of 50 to 1200 US gal/min (3 to 76 L/s), usually with nozzle diameters in the range of 0.5 to 1.9 inches (10 to 50 mm). Guns are used not only for irrigation, but also for industrial applications such as dust suppression and logging.

Sprinklers can also be mounted on moving platforms connected to the water source by a hose. Automatically moving wheeled systems known as *traveling sprinklers* may irrigate areas such as small farms, sports fields, parks, pastures, and cemeteries unattended. Most of these utilize a length of polyethylene tubing wound on a steel drum. As the tubing is wound on the drum powered by the irrigation water or a small gas engine, the sprinkler is pulled across the field. When the sprinkler arrives back at the reel the system shuts off. This type of system is known to most people as a "waterreel" traveling irrigation sprinkler and they are used extensively for dust suppression, irrigation, and land application of waste water.

Other travelers use a flat rubber hose that is dragged along behind while the sprinkler platform is pulled by a cable. These cable-type travelers are definitely old technology and their use is limited in today's modern irrigation projects.

Irrigation using Center Pivot

A small center pivot system from beginning to end

Wheel line irrigation system in Idaho, 2001

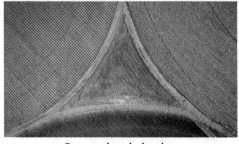

Center pivot irrigation

Center pivot irrigation is a form of sprinkler irrigation consisting of several segments of pipe (usually galvanized steel or aluminium) joined together and supported by trusses, mounted on wheeled towers with sprinklers positioned along its length. The system moves in a circular pattern and is fed with water from the pivot point at the center of the arc. These systems are found and used in all parts of the world and allow irrigation of all types of terrain. Newer systems have drop sprinkler heads as shown in the image that follows.

Most center pivot systems now have drops hanging from a u-shaped pipe attached at the top of the pipe with sprinkler head that are positioned a few feet (at most) above the crop, thus limiting evaporative losses. Drops can also be used with drag hoses or bubblers that deposit the water directly on the ground between crops. Crops are often planted in a circle to conform to the center pivot. This type of system is known as LEPA (Low Energy Precision Application). Originally, most center pivots were water powered. These were replaced by hydraulic systems (*T-L Irrigation*) and electric motor driven systems (Reinke, Valley, Zimmatic). Many modern pivots feature GPS devices.

Irrigation by Lateral Move (Side Roll, Wheel Line, Wheelmove)

A *series of pipes, each with a wheel* of about 1.5 m diameter permanently affixed to its midpoint, and sprinklers along its length, are coupled together. Water is supplied at one end using a large hose. After sufficient irrigation has been applied to one strip of the field, the hose is removed, the water drained from the system, and the assembly rolled either by hand or with a purpose-built mechanism, so that the sprinklers are moved to a different position across the field. The hose is reconnected. The process is repeated in a pattern until the whole field has been irrigated.

This system is less expensive to install than a center pivot, but much more labor-intensive to operate - it does not travel automatically across the field: it applies water in a stationary strip, must be drained, and then rolled to a new strip. Most systems use 4 or 5-inch (130 mm) diameter aluminum pipe. The pipe doubles both as water transport and as an axle for rotating all the wheels. A drive system (often found near the centre of the wheel line) rotates the clamped-together pipe sections as a single axle, rolling the whole wheel line. Manual adjustment of individual wheel positions may be necessary if the system becomes misaligned.

Wheel line systems are limited in the amount of water they can carry, and limited in the height of crops that can be irrigated. One useful feature of a lateral move system is that it consists of sections that can be easily disconnected, adapting to field shape as the line is moved. They are most often used for small, rectilinear, or oddly-shaped fields, hilly or mountainous regions, or in regions where labor is inexpensive.

Sub-irrigation

Subirrigation has been used for many years in field crops in areas with high water tables. It is a method of artificially raising the water table to allow the soil to be moistened from below the plants' root zone. Often those systems are located on permanent grasslands in lowlands or river valleys and combined with drainage infrastructure. A system of pumping stations, canals, weirs and gates allows it to increase or decrease the water level in a network of ditches and thereby control the water table.

Sub-irrigation is also used in commercial greenhouse production, usually for potted plants. Water is delivered from below, absorbed upwards, and the excess collected for recycling. Typically, a solution of water and nutrients floods a container or flows through a trough for a short period of time, 10–20 minutes, and is then pumped back into a holding tank for reuse. Sub-irrigation in greenhouses requires fairly sophisticated, expensive equipment and management. Advantages are water and nutrient conservation, and labor-saving through lowered system maintenance and automation. It is similar in principle and action to subsurface basin irrigation.

Irrigation Automatically, Non-electric using Buckets and Ropes

Besides the common manual watering by bucket, an automated, natural version of this also exists. Using plain polyester ropes combined with a prepared ground mixture can be used to water plants from a vessel filled with water.

The ground mixture would need to be made depending on the plant itself, yet would mostly consist of black potting soil, vermiculite and perlite. This system would (with certain crops) allow to save expenses as it does not consume any electricity and only little water (unlike sprinklers, water timers, etc.). However, it may only be used with certain crops (probably mostly larger crops that do not need a humid environment; perhaps e.g. paprikas).

Irrigation using Water Condensed from Humid Air

In countries where at night, humid air sweeps the countryside.Water can be obtained from the humid air by condensation onto cold surfaces. This is for example practiced in the vineyards at Lanzarote using stones to condense water or with various fog collectors based on canvas or foil sheets.

In-ground Irrigation

Most commercial and residential irrigation systems are "in ground" systems, which means that everything is buried in the ground. With the pipes, sprinklers, emitters (drippers), and irrigation valves being hidden, it makes for a cleaner, more presentable landscape without garden hoses or other items having to be moved around manually. This does, however, create some drawbacks in the maintenance of a completely buried system.

Most irrigation systems are divided into zones. A zone is a single irrigation valve and one or a group of drippers or sprinklers that are connected by pipes or tubes. Irrigation systems are divided into zones because there is usually not enough pressure and available flow to run sprinklers for an entire yard or sports field at once. Each zone has a solenoid valve on it that is controlled via wire by an irrigation controller. The irrigation controller is either a mechanical (now the "dinosaur" type) or electrical device that signals a zone to turn on at a specific time and keeps it on for a specified amount of time. "Smart Controller" is a recent term for a controller that is capable of adjusting the watering time by itself in response to current environmental conditions. The smart controller determines current conditions by means of historic weather data for the local area, a soil moisture sensor (water potential or water content), rain sensor, or in more sophisticated systems satellite feed weather station, or a combination of these.

When a zone comes on, the water flows through the lateral lines and ultimately ends up at the irrigation emitter (drip) or sprinkler heads. Many sprinklers have pipe thread inlets on the bottom of them which allows a fitting and the pipe to be attached to them. The sprinklers are usually installed with the top of the head flush with the ground surface. When the water is pressurized, the head will pop up out of the ground and water the desired area until the valve closes and shuts off that zone. Once there is no more water pressure in the lateral line, the sprinkler head will retract back into the ground. Emitters are generally laid on the soil surface or buried a few inches to reduce evaporation losses.

Water Sources

Irrigation is underway by pump-enabled extraction directly from the Gumti, seen in the background, in Comilla, Bangladesh.

Irrigation water can come from groundwater (extracted from springs or by using wells), from surface water (withdrawn from rivers, lakes or reservoirs) or from non-conventional sources like treated wastewater, desalinated water or drainage water. A special form of irrigation using surface water is spate irrigation, also called floodwater harvesting. In case of a flood (spate), water is diverted to normally dry river beds (wadis) using a network of dams, gates and channels and spread over large areas. The moisture stored in the soil will be used thereafter to grow crops. Spate irrigation areas are in particular located in semi-arid or arid, mountainous regions. While floodwater harvesting belongs to the accepted irrigation methods, rainwater harvesting is usually not considered as a form of irrigation. Rainwater harvesting is the collection of runoff water from roofs or unused land and the concentration of this.

Around 90% of wastewater produced globally remains untreated, causing widespread water pollution, especially in low-income countries. Increasingly, agriculture uses untreated wastewater as a source of irrigation water. Cities provide lucrative markets for fresh produce, so are attractive to farmers. However, because agriculture has to compete for increasingly scarce water resources with industry and municipal users, there is often no alternative for farmers but to use water polluted with urban waste, including sewage, directly to water their crops. Significant health hazards can result from using water loaded with pathogens in this way, especially if people eat raw vegetables that have been irrigated with the polluted water. The International Water Management Institute has worked in India, Pakistan, Vietnam, Ghana, Ethiopia, Mexico and other countries on various projects aimed at assessing and reducing risks of wastewater irrigation. They advocate a 'multiple-barrier' approach to wastewater use, where farmers are encouraged to adopt various risk-reducing behaviours. These include ceasing irrigation a few days before harvesting to allow pathogens to die off in the sunlight, applying water carefully so it does not contaminate leaves likely to be eaten raw, cleaning vegetables with disinfectant or allowing fecal sludge used in farming to dry before being used as a human manure. The World Health Organization has developed guidelines for safe water use.

There are numerous benefits of using recycled water for irrigation, including the low cost (when compared to other sources, particularly in an urban area), consistency of supply (regardless of season, climatic conditions and associated water restrictions), and general consistency of quality. Irrigation of recycled wastewater is also considered as a means for plant fertilization and particularly nutrient supplementation. This approach carries with it a risk of soil and water pollution through excessive wastewater application. Hence, a detailed understanding of soil water conditions is essential for effective utilization of wastewater for irrigation.

Efficiency

Young engineers restoring and developing the old Mughal irrigation system during the reign of the Mughal Emperor Bahadur Shah II

Modern irrigation methods are efficient enough to supply the entire field uniformly with water, so that each plant has the amount of water it needs, neither too much nor too little. Water use efficiency in the field can be determined as follows:

- Field Water Efficiency (%) = (Water Transpired by Crop ÷ Water Applied to Field) x 100

Until 1960s, the common perception was that water was an infinite resource. At that time, there were fewer than half the current number of people on the planet. People were not as wealthy as today, consumed fewer calories and ate less meat, so less water was needed to produce their food. They required a third of the volume of water we presently take from rivers. Today, the competition for water resources is much more intense. This is because there are now more than seven billion people on the planet, their consumption of water-thirsty meat and vegetables is rising, and there is increasing competition for water from industry, urbanisation and biofuel crops. To avoid a global water crisis, farmers will have to strive to increase productivity to meet growing demands for food, while industry and cities find ways to use water more efficiently.

Successful agriculture is dependent upon farmers having sufficient access to water. However, water scarcity is already a critical constraint to farming in many parts of the world. With regards to agriculture, the World Bank targets food production and water management as an increasingly global issue that is fostering a growing debate. Physical water scarcity is where there is not enough water to meet all demands, including that needed for ecosystems to function effectively. Arid regions frequently suffer from physical water scarcity. It also occurs where water seems abundant but where resources are over-committed. This can happen where there is overdevelopment of hydraulic infrastructure, usually for irrigation. Symptoms of physical water scarcity include environmental degradation and declining groundwater. Economic scarcity, meanwhile, is caused by a lack of investment in water or insufficient human capacity to satisfy the demand for water. Symptoms of economic water scarcity include a lack of infrastructure, with people often having to fetch water from rivers for domestic and agricultural uses. Some 2.8 billion people currently live in water-scarce areas.

Technical Challenges

Irrigation schemes involve solving numerous engineering and economic problems while minimizing negative environmental impact.

- Competition for surface water rights.

- Overdrafting (depletion) of underground aquifers.

- Ground subsidence (e.g. New Orleans, Louisiana)

- Underirrigation or irrigation giving only just enough water for the plant (e.g. in drip line irrigation) gives poor soil salinity control which leads to increased soil salinity with consequent buildup of toxic salts on soil surface in areas with high evaporation. This requires either leaching to remove these salts and a method of drainage to carry the salts away. When using drip lines, the leaching is best done regularly at certain intervals (with only a slight excess of water), so that the salt is flushed back under the plant's roots.

- Overirrigation because of poor distribution uniformity or management wastes water, chemicals, and may lead to water pollution.

- Deep drainage (from over-irrigation) may result in rising water tables which in some instances will lead to problems of irrigation salinity requiring watertable control by some form of subsurface land drainage.

- Irrigation with saline or high-sodium water may damage soil structure owing to the formation of alkaline soil

- Clogging of filters: It is mostly algae that clog filters, drip installations and nozzles. UV and ultrasonic method can be used for algae control in irrigation systems.

Soil Salinity

Visibly salt-affected soils on rangeland in Colorado. Salts dissolved from the soil accumulate at the soil surface and are deposited on the ground and at the base of the fence post.

Soil salinity is the salt content in the soil; the process of increasing the salt content is known as salinization. Salts occur naturally within soils and water. Salination can be caused by natural processes such as mineral weathering or by the gradual withdrawal of an ocean. It can also come about through artificial processes such as irrigation.

Natural Occurrence

Salts are a natural component in soils and water. The ions responsible for salination are: Na^+, K^+, Ca^{2+}, Mg^{2+} and Cl^-. As the Na^+ (sodium) predominates, soils can become sodic. Sodic soils present particular challenges because they tend to have very poor structure which limits or prevents water infiltration and drainage.

Over long periods of time, as soil minerals weather and release salts, these salts are flushed or leached out of the soil by drainage water in areas with sufficient precipitation. In addition to mineral weathering, salts are also deposited via dust and precipitation. In dry regions salts may accumulate, leading to naturally saline soils. This is the case, for example, in large parts of Australia. Human practices can increase the salinity of soils by the addition of salts in irrigation water. Proper irrigation management can prevent salt accumulation by providing adequate drainage water to leach added salts from the soil. Disrupting drainage patterns that provide leaching can also result in salt accumulations. An example of this occurred in Egypt in 1970 when the Aswan High Dam was built. The change in the level of ground water before the construction had enabled soil erosion, which led to high concentration of salts in the water table. After the construction, the continuous high level of the water table led to the salination of the arable land.

Dry Land Salinity

Salinity in drylands can occur when the water table is between two and three metres from the surface of the soil. The salts from the groundwater are raised by capillary action to the surface of the soil. This occurs when groundwater is saline (which is true in many areas), and is favored by land use practices allowing more rainwater to enter the aquifer than it could accommodate. For example, the clearing of trees for agriculture is a major reason for dryland salinity in some areas, since deep rooting of trees has been replaced by shallow rooting of annual crops.

Salinity Due to Irrigation

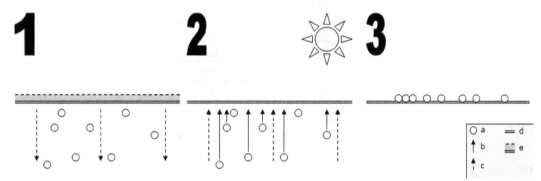

Rain or irrigation, in the absence of leaching, can bring salts to the surface by capillary action

Salinity from irrigation can occur over time wherever irrigation occurs, since almost all water (even natural rainfall) contains some dissolved salts. When the plants use the water, the salts are left behind in the soil and eventually begin to accumulate. Since soil salinity makes it more difficult for plants to absorb soil moisture, these salts must be leached out of the plant root zone by applying additional water. This water in excess of plant needs is called the leaching fraction. Salination from irrigation water is also greatly increased by poor drainage and use of saline water for irrigating agricultural crops.

Salinity in urban areas often results from the combination of irrigation and groundwater processes. Irrigation is also now common in cities (gardens and recreation areas).

Consequences of Salinity

The consequences of salinity are

- detrimental effects on plant growth and yield

- damage to infrastructure (roads, bricks, corrosion of pipes and cables)

- reduction of water quality for users, sedimentation problems

- soil erosion ultimately, when crops are too strongly affected by the amounts of salts.

Salinity is an important land degradation problem. Soil salinity can be reduced by leaching soluble salts out of soil with excess irrigation water. Soil salinity control involves watertable control and flushing in combination with tile drainage or another form of subsurface drainage. A comprehensive treatment of soil salinity is available from the United Nations Food and Agriculture Organization.

High levels of soil salinity can be tolerated if salt-tolerant plants are grown. Sensitive crops lose their vigor already in slightly saline soils, most crops are negatively affected by (moderately) saline soils, and only salinity resistant crops thrive in severely saline soils. The University of Wyoming and the Government of Alberta report data on the salt tolerance of plants.

Regions Affected

From the FAO/UNESCO Soil Map of the World the following salinised areas can be derived.

Region	Area (10⁶ ha)
Africa	69.5
Near and Middle East	53.1
Asia and Far East	19.5
Latin America	59.4
Australia	84.7
North America	16.0

Aquifer

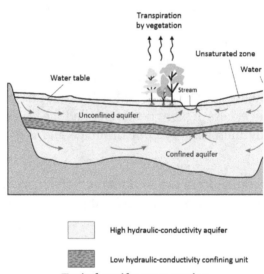

Typical aquifer cross-section

An aquifer is an underground layer of water-bearing permeable rock, rock fractures or unconsolidated materials (gravel, sand, or silt) from which groundwater can be extracted using a water well. The study of water flow in aquifers and the characterization of aquifers is called hydrogeology. Related terms include aquitard, which is a bed of low permeability along an aquifer, and aquiclude (or *aquifuge*), which is a solid, impermeable area underlying or overlying an aquifer. If the impermeable area overlies the aquifer, pressure could cause it to become a confined aquifer.

Depth

Aquifers may occur at various depths. Those closer to the surface are not only more likely to be used for water supply and irrigation, but are also more likely to be topped up by the local rainfall. Many desert areas have limestone hills or mountains within them or close to them that can be exploited as groundwater resources. Part of the Atlas Mountains in North Africa, the Lebanon and Anti-Lebanon ranges between Syria and Lebanon, the Jebel Akhdar (Oman) in Oman, parts of the Sierra Nevada and neighboring ranges in the United States' Southwest, have shallow aquifers that are exploited for their water. Overexploitation can lead to the exceeding of the practical sustained yield; i.e., more water is taken out than can be replenished. Along the coastlines of certain countries, such as Libya and Israel, increased water usage associated with population growth has caused a lowering of the water table and the subsequent contamination of the groundwater with saltwater from the sea.

The beach provides a model to help visualize an aquifer. If a hole is dug into the sand, very wet or saturated sand will be located at a shallow depth. This hole is a crude well, the wet sand represents an aquifer, and the level to which the water rises in this hole represents the water table.

In 2013 large freshwater aquifers were discovered under continental shelves off Australia, China, North America and South Africa. They contain an estimated half a million cubic kilometers of "low salinity" water that could be economically processed into potable water. The reserves formed when ocean levels were lower and rainwater made its way into the ground in land areas that were not submerged until the ice age ended 20,000 years ago. The volume is estimated to be 100x the amount of water extracted from other aquifers since 1900.

Classification

The above diagram indicates typical flow directions in a cross-sectional view of a simple confined or unconfined aquifer system. The system shows two aquifers with one aquitard (a confining or imper-meable layer) between them, surrounded by the bedrock *aquiclude*, which is in contact with a gaining stream (typical in humid regions). The water table and unsaturated zone are also illustrated. An *aqui-tard* is a zone within the earth that restricts the flow of groundwater from one aquifer to another. An aquitard can sometimes, if completely impermeable, be called an *aquiclude* or *aquifuge*. Aquitards are composed of layers of either clay or non-porous rock with low hydraulic conductivity.

Saturated Versus Unsaturated

Groundwater can be found at nearly every point in the Earth's shallow subsurface to some degree, although aquifers do not necessarily contain fresh water. The Earth's crust can be divided into two regions: the *saturated zone* or *phreatic zone* (e.g., aquifers, aquitards, etc.), where all available spaces are filled with water, and the *unsaturated zone* (also called the vadose zone), where there are still pockets of air that contain some water, but can be filled with more water.

Saturated means the pressure head of the water is greater than atmospheric pressure (it has a gauge pressure > 0). The definition of the water table is the surface where the pressure head is equal to atmospheric pressure (where gauge pressure = 0).

Unsaturated conditions occur above the water table where the pressure head is negative (absolute pressure can never be negative, but gauge pressure can) and the water that incompletely fills the pores of the aquifer material is under suction. The water content in the unsaturated zone is held in place by surface adhesive forces and it rises above the water table (the zero-gauge-pressure isobar) by capillary action to saturate a small zone above the phreatic surface (the capillary fringe) at less than atmospheric pressure. This is termed tension saturation and is not the same as saturation on a water-content basis. Water content in a capillary fringe decreases with increasing distance from the phreatic surface. The capillary head depends on soil pore size. In sandy soils with larger pores, the head will be less than in clay soils with very small pores. The normal capillary rise in a clayey soil is less than 1.80 m (six feet) but can range between 0.3 and 10 m (one and 30 ft).

The capillary rise of water in a small-diameter tube involves the same physical process. The water table is the level to which water will rise in a large-diameter pipe (e.g., a well) that goes down into the aquifer and is open to the atmosphere.

Aquifers Versus Aquitards

Aquifers are typically saturated regions of the subsurface that produce an economically feasible quantity of water to a well or spring (e.g., sand and gravel or fractured bedrock often make good aquifer materials).

An aquitard is a zone within the earth that restricts the flow of groundwater from one aquifer to another. A completely impermeable aquitard is called an aquiclude or aquifuge. Aquitards comprise layers of either clay or non-porous rock with low hydraulic conductivity.

In mountainous areas (or near rivers in mountainous areas), the main aquifers are typically unconsolidated alluvium, composed of mostly horizontal layers of materials deposited by water processes (rivers and streams), which in cross-section (looking at a two-dimensional slice of the aquifer) appear to be layers of alternating coarse and fine materials. Coarse materials, because of the high energy needed to move them, tend to be found nearer the source (mountain fronts or rivers), whereas the fine-grained material will make it farther from the source (to the flatter parts of the basin or overbank areas - sometimes called the pressure area). Since there are less fine-grained deposits near the source, this is a place where aquifers are often unconfined (sometimes called the forebay area), or in hydraulic communication with the land surface.

Confined Versus Unconfined

There are two end members in the spectrum of types of aquifers; *confined* and *unconfined* (with semi-confined being in between). Unconfined aquifers are sometimes also called *water table* or *phreatic* aquifers, because their upper boundary is the water table or phreatic surface. Typically (but not always) the shallowest aquifer at a given location is unconfined, meaning it does not have a confining layer (an aquitard or aquiclude) between it and the surface. The term "perched" refers to ground water accumulating above a low-permeability unit or strata, such as a clay layer. This term is generally used to refer to a small local area of ground water that occurs at an elevation higher than a regionally extensive aquifer. The difference between perched and unconfined aquifers is their size (perched is smaller). Confined aquifers are aquifers that are overlain by a confining layer, often made up of clay. The confining layer might offer some protection from surface contamination.

If the distinction between confined and unconfined is not clear geologically (i.e., if it is not known if a clear confining layer exists, or if the geology is more complex, e.g., a fractured bedrock aquifer), the value of storativity returned from an aquifer test can be used to determine it (although aquifer tests in unconfined aquifers should be interpreted differently than confined ones). Confined aquifers have very low storativity values (much less than 0.01, and as little as 10^{-5}), which means that the aquifer is storing water using the mechanisms of aquifer matrix expansion and the compressibility of water, which typically are both quite small quantities. Unconfined aquifers have storativities (typically then called specific yield) greater than 0.01 (1% of bulk volume); they release water from storage by the mechanism of actually draining the pores of the aquifer, releasing relatively large amounts of water (up to the drainable porosity of the aquifer material, or the minimum volumetric water content).

Isotropic Versus Anisotropic

In isotropic aquifers or aquifer layers the hydraulic conductivity (K) is equal for flow in all directions, while in anisotropic conditions it differs, notably in horizontal (Kh) and vertical (Kv) sense.

Semi-confined aquifers with one or more aquitards work as an anisotropic system, even when the separate layers are isotropic, because the compound Kh and Kv values are different .

When calculating flow to drains or flow to wells in an aquifer, the anisotropy is to be taken into account lest the resulting design of the drainage system may be faulty.

Groundwater in Rock Formations

Groundwater may exist in *underground rivers* (e.g., caves where water flows freely underground). This may occur in eroded limestone areas known as karst topography, which make up only a small percentage of Earth's area. More usual is that the pore spaces of rocks in the subsurface are simply saturated with water — like a kitchen sponge — which can be pumped out for agricultural, industrial, or municipal uses.

If a rock unit of low porosity is highly fractured, it can also make a good aquifer (via fissure flow), provided the rock has a hydraulic conductivity sufficient to facilitate movement of water. Porosity is important, but, *alone*, it does not determine a rock's ability to act as an aquifer. Areas of the Deccan Traps (a basaltic lava) in west central India are good examples of rock formations with high porosity but low permeability, which makes them poor aquifers. Similarly, the micro-porous (Upper Cretaceous) Chalk of south east England, although having a reasonably high porosity, has a low grain-to-grain permeability, with its good water-yielding characteristics mostly due to micro-fracturing and fissuring.

Human Dependence on Groundwater

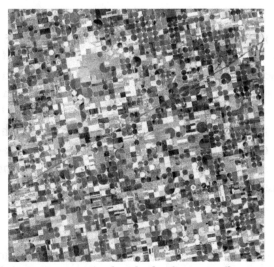

Center-pivot irrigated fields in Kansas covering hundreds of square miles watered by the Ogallala Aquifer

Most land areas on Earth have some form of aquifer underlying them, sometimes at significant depths. In some cases, these aquifers are rapidly being depleted by the human population.

Fresh-water aquifers, especially those with limited recharge by snow or rain, also known as meteoric water, can be over-exploited and depending on the local hydrogeology, may draw in non-potable water or saltwater intrusion from hydraulically connected aquifers or surface water bodies. This can be a serious problem, especially in coastal areas and other areas where aquifer pumping is excessive. In some areas, the ground water can become contaminated by arsenic and other mineral poisons.

Aquifers are critically important in human habitation and agriculture. Deep aquifers in arid areas have long been water sources for irrigation. Many villages and even large cities draw their water supply from wells in aquifers.

Municipal, irrigation, and industrial water supplies are provided through large wells. Multiple wells for one water supply source are termed "wellfields", which may withdraw water from confined or unconfined aquifers. Using ground water from deep, confined aquifers provides more protection from surface water contamination. Some wells, termed "collector wells," are specifically designed to induce infiltration of surface (usually river) water.

Aquifers that provide sustainable fresh groundwater to urban areas and for agricultural irrigation are typically close to the ground surface (within a couple of hundred metres) and have some recharge by fresh water. This recharge is typically from rivers or meteoric water (precipitation) that percolates into the aquifer through overlying unsaturated materials.

Occasionally, sedimentary or "fossil" aquifers are used to provide irrigation and drinking water to urban areas. In Libya, for example, Muammar Gaddafi's Great Manmade River project has pumped large amounts of groundwater from aquifers beneath the Sahara to populous areas near the coast. Though this has saved Libya money over the alternative, desalination, the aquifers are likely to run dry in 60 to 100 years. Aquifer depletion has been cited as one of the causes of the food price rises of 2011.

Subsidence

In unconsolidated aquifers, groundwater is produced from pore spaces between particles of gravel, sand, and silt. If the aquifer is confined by low-permeability layers, the reduced water pressure in the sand and gravel causes slow drainage of water from the adjoining confining layers. If these confining layers are composed of compressible silt or clay, the loss of water to the aquifer reduces the water pressure in the confining layer, causing it to compress from the weight of overlying geologic materials. In severe cases, this compression can be observed on the ground surface as subsidence. Unfortunately, much of the subsidence from groundwater extraction is permanent (elastic rebound is small). Thus, the subsidence is not only permanent, but the compressed aquifer has a permanently reduced capacity to hold water.

Saltwater Intrusion

Aquifers near the coast have a lens of freshwater near the surface and denser seawater under freshwater. Seawater penetrates the aquifer diffusing in from the ocean and is denser than freshwater.

For porous (i.e., sandy) aquifers near the coast, the thickness of freshwater atop saltwater is about 40 feet (12 m) for every 1 ft (0.30 m) of freshwater head above sea level. This relationship is called the Ghyben-Herzberg equation. If too much ground water is pumped near the coast, salt-water may intrude into freshwater aquifers causing contamination of potable freshwater supplies. Many coastal aquifers, such as the Biscayne Aquifer near Miami and the New Jersey Coastal Plain aquifer, have problems with saltwater intrusion as a result of overpumping and sea level rise.

Salination

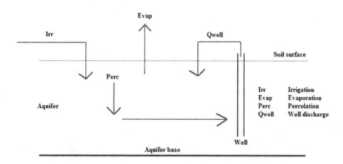

Water balance in the aquifer of a surface irrigated area with reuse of percolation water by pumping from wells

Diagram of a water balance of the aquifer

Aquifers in surface irrigated areas in semi-arid zones with reuse of the unavoidable irrigation water losses percolating down into the underground by supplemental irrigation from wells run the risk of salination.

Surface irrigation water normally contains salts in the order of 0.5 g/l or more and the annual irrigation requirement is in the order of 10000 m³/ha or more so the annual import of salt is in the order of 5000 kg/ha or more.

Under the influence of continuous evaporation, the salt concentration of the aquifer water may increase continually and eventually cause an environmental problem.

For salinity control in such a case, annually an amount of drainage water is to be discharged from the aquifer by means of a subsurface drainage system and disposed of through a safe outlet. The drainage system may be *horizontal* (i.e. using pipes, tile drains or ditches) or *vertical* (drainage by wells). To estimate the drainage requirement, the use of a groundwater model with an agro-hydro-salinity component may be instrumental, e.g. SahysMod.

Examples

The Great Artesian Basin situated in Australia is arguably the largest groundwater aquifer in the world (over 1.7 million km²). It plays a large part in water supplies for Queensland and remote parts of South Australia.

The Guarani Aquifer, located beneath the surface of Argentina, Brazil, Paraguay, and Uruguay, is one of the world's largest aquifer systems and is an important source of fresh water. Named after the Guarani people, it covers 1,200,000 km², with a volume of about 40,000 km³, a thickness of between 50 m and 800 m and a maximum depth of about 1,800 m.

Aquifer depletion is a problem in some areas, and is especially critical in northern Africa. However, new methods of groundwater management such as artificial recharge and injection of surface waters during seasonal wet periods has extended the life of many freshwater aquifers, especially in the United States.

The Ogallala Aquifer of the central United States is one of the world's great aquifers, but in places it is being rapidly depleted by growing municipal use, and continuing agricultural use. This huge aquifer, which underlies portions of eight states, contains primarily fossil water from the time of the last glaciation. Annual recharge, in the more arid parts of the aquifer, is estimated to total only about 10 percent of annual withdrawals. According to a 2013 report by research hydrologist Leonard F. Konikow at the United States Geological Survey (USGC), the depletion between 2001–2008, inclusive, is about 32 percent of the cumulative depletion during the entire 20th century (Konikow 2013:22)." In the United States, the biggest users of water from aquifers include agricultural irrigation and oil and coal extraction. "Cumulative total groundwater depletion in the United States accelerated in the late 1940s and continued at an almost steady linear rate through the end of the century. In addition to widely recognized environmental consequences, groundwater depletion also adversely impacts the long-term sustainability of groundwater supplies to help meet the Nation's water needs."

An example of a significant and sustainable carbonate aquifer is the Edwards Aquifer in central Texas. This carbonate aquifer has historically been providing high quality water for nearly 2 million people, and even today, is full because of tremendous recharge from a number of area streams, rivers and lakes. The primary risk to this resource is human development over the recharge areas.

Discontinuous sand bodies at the base of the McMurray Formation in the Athabasca Oil Sands region of northeastern Alberta, Canada, are commonly referred to as the Basal Water Sand (BWS) aquifers. Saturated with water, they are confined beneath impermeable bitumen-saturated sands that are exploited to recover bitumen for synthetic crude oil production. Where they are deep-lying and recharge occurs from underlying Devonian formations they are saline, and where they are shallow and recharged by meteoric water they are non-saline. The BWS typically pose problems for the recovery of bitumen, whether by open-pit mining or by *in situ* methods such as steam-assisted gravity drainage (SAGD), and in some areas they are targets for waste-water injection.

Seawater

Seawater in the Strait of Malacca

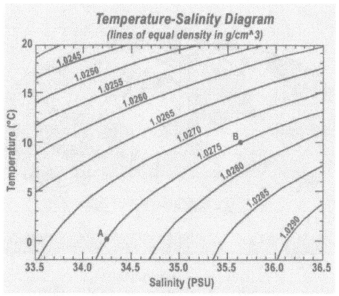

Temperature-salinity diagram of changes in density of water

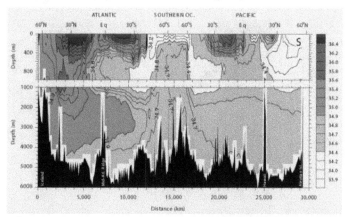

Changes in ocean water density with depth at different latitudes

Seawater, or salt water, is water from a sea or ocean. On average, seawater in the world's oceans has a salinity of about 3.5% (35 g/L, or 600 mM). This means that every kilogram (roughly one litre by volume) of seawater has approximately 35 grams (1.2 oz) of dissolved salts (predominantly sodium (Na+) and chloride (Cl–) ions). Average density at the surface is 1.025 kg/l. Seawater is denser than both fresh water and pure water (density 1.0 kg/l at 4 °C (39 °F)) because the dissolved salts increase the mass by a larger proportion than the volume. The freezing point of seawater decreases as salt concentration increases. At typical salinity, it freezes at about –2 °C (28 °F). The coldest seawater ever recorded (in a liquid state) was in 2010, in a stream under an Antarctic glacier, and measured –2.6 °C (27.3 °F). Seawater pH is typically limited to a range between 7.5 and 8.4. However, there is no universally accepted reference pH-scale for seawater and the difference between measurements based on different reference scales may be up to 0.14 units.

Geochemistry

The thermal conductivity of seawater is 0.6 W/mK at 25 °C and a salinity of 35 g/kg. The thermal conductivity decreases with increasing salinity and increases with increasing temperature.

Salinity

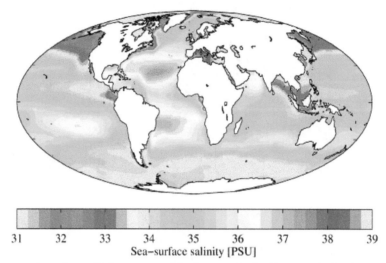

Annual mean sea surface salinity expressed in the Practical Salinity Scale for the World Ocean.
Data from the World Ocean Atlas

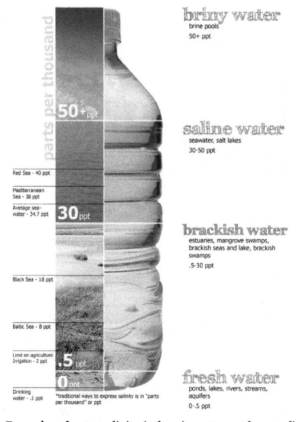

Examples of water salinity (values in parts per thousand)

Although the vast majority of seawater has a salinity of between 3.1% and 3.8%, seawater is not uniformly saline throughout the world. Where mixing occurs with fresh water runoff from river mouths or near melting glaciers, seawater can be substantially less saline. The most saline open sea is the Red Sea, where high rates of evaporation, low precipitation and river inflow, and confined circulation result in unusually salty water. The salinity in isolated bodies of water (for example, the Dead Sea) can be considerably greater still.

The density of surface seawater ranges from about 1020 to 1029 kg/m³, depending on the temperature and salinity. Deep in the ocean, under high pressure, seawater can reach a density of 1050 kg/m³ or higher. Seawater pH is limited to the range 7.5 to 8.4. The speed of sound in seawater is about 1,500 m/s, and varies with water temperature, salinity, and pressure.

Compositional Differences from Freshwater

Seawater composition (by mass) (salinity = 3.5%)			
Element	**Percent**	**Element**	**Percent**
Oxygen	85.84	Sulfur	0.091
Hydrogen	10.82	Calcium	0.04
Chloride	1.94	Potassium	0.04
Sodium	1.08	Bromide	0.0067
Magnesium	0.1292	Carbon	0.0028
Vanadium	1.5×10^{-11} - 3.3×10^{-11}		

Seawater contains more dissolved ions than all types of freshwater. However, the ratios of solutes differ dramatically. For instance, although seawater contains about 2.8 times more bicarbonate than river water based on molarity, the percentage of bicarbonate in seawater as a ratio of *all* dissolved ions is far lower than in river water. Bicarbonate ions also constitute 48% of river water solutes but only 0.14% of all seawater ions. Differences like these are due to the varying residence times of seawater solutes; sodium and chlorine have very long residence times, while calcium (vital for carbonate formation) tends to precipitate much more quickly. The most abundant dissolved ions in seawater are sodium, chloride, magnesium, sulfate and calcium. Its osmolarity is about 1000 mOsm/l.

Small amounts of other substances are found including amino acids at concentrations up to 2 micrograms of Nitrogen atoms per liter, which are thought to have played a key role in the origin of life.

Microbial Components

Research in 1957 by the Scripps Institution of Oceanography sampled water in both pelagic and neritic locations in the Pacific Ocean. Direct microscopic counts and cultures were used, the direct counts in some cases showing up to 10 000 times that obtained from cultures. These differences were attributed to the occurrence of bacteria in aggregates, selective effects of the culture media, and the presence of inactive cells. A marked reduction in bacterial culture numbers was noted below the thermocline, but not by direct microscopic observation. Large numbers of spirilli-like forms were seen by microscope but not under cultivation. The disparity in numbers obtained by the two methods is well known in this and other fields. In the 1990s, improved techniques of detection and identification of microbes by probing just small snippets of DNA, enabled researchers taking part in the Census of Marine Life to identify thousands of previously unknown microbes usually present only in small numbers. This revealed a far greater diversity than previously suspected, so that a litre of seawater may hold more than 20,000 species. Dr. Mitchell Sogin from the

Marine Biological Laboratory feels that "the number of different kinds of bacteria in the oceans could eclipse five to 10 million."

Bacteria are found at all depths in the water column, as well as in the sediments, some being aerobic, others anaerobic. Most are free-swimming, but some exist as symbionts within other organisms – examples of these being bioluminescent bacteria. Cyanobacteria played an important role in the evolution of ocean processes, enabling the development of stromatolites and oxygen in the atmosphere.

Some bacteria interact with diatoms, and form a critical link in the cycling of silicon in the ocean. One anaerobic species, *Thiomargarita namibiensis*, plays an important part in the breakdown of hydrogen sulphide eruptions from diatomaceous sediments off the Namibian coast, and generated by high rates of phytoplankton growth in the Benguela Current upwelling zone, eventually falling to the seafloor.

Bacteria-like Archaea surprised marine microbiologists by their survival and thriving in extreme environments, such as the hydrothermal vents on the ocean floor. Alkalotolerant marine bacteria such as *Pseudomonas* and *Vibrio* spp. survive in a pH range of 7.3 to 10.6, while some species will grow only at pH 10 to 10.6. Archaea also exist in pelagic waters and may constitute as much as half the ocean's biomass, clearly playing an important part in oceanic processes. In 2000 sediments from the ocean floor revealed a species of Archaea that breaks down methane, an important greenhouse gas and a major contributor to atmospheric warming. Some bacteria break down the rocks of the sea floor, influencing seawater chemistry. Oil spills, and runoff containing human sewage and chemical pollutants have a marked effect on microbial life in the vicinity, as well as harbouring pathogens and toxins affecting all forms of marine life. The protist dinoflagellates may at certain times undergo population explosions called blooms or red tides, often after human-caused pollution. The process may produce metabolites known as biotoxins, which move along the ocean food chain, tainting higher-order animal consumers.

Pandoravirus salinus, a species of very large virus, with a genome much larger than that of any other virus species, was discovered in 2013. Like the other very large viruses *Mimivirus* and *Megavirus*, *Pandoravirus* infects amoebas, but its genome, containing 1.9 to 2.5 megabases of DNA, is twice as large as that of *Megavirus*, and it differs greatly from the other large viruses in appearance and in genome structure.

In 2013 researchers from Aberdeen University announced that they were starting a hunt for undiscovered chemicals in organisms that have evolved in deep sea trenches, hoping to find "the next generation" of antibiotics, anticipating an "antibiotic apocalypse" with a dearth of new infection-fighting drugs. The EU-funded research will start in the Atacama Trench and then move on to search trenches off New Zealand and Antarctica.

The ocean has a long history of human waste disposal on the assumption that its vast size makes it capable of absorbing and diluting all noxious material. While this may be true on a small scale, the large amounts of sewage routinely dumped has damaged many coastal ecosystems, and rendered them life-threatening. Pathogenic viruses and bacteria occur in such waters, such as *Escherichia coli*, *Vibrio cholerae* the cause of cholera, hepatitis A, hepatitis E and polio, along with protozoans causing giardiasis and cryptosporidiosis. These pathogens are routinely present in the ballast water of large vessels, and are widely spread when the ballast is discharged.

Origin

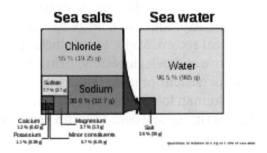

Diagram showing concentrations of various salt ions in seawater. The composition of the total salt component is: Cl–55%, Na+30.6%, SO2–4 7.7%, Mg2+3.7%, Ca2+1.2%, K+1.1%, Other 0.7%. Note that the diagram is only correct when in units of wt/wt, not wt/vol or vol/vol.

Total Molar Composition of Seawater (Salinity = 35)	
Component	Concentration (mol/kg)
H$_2$O	53.6
Cl–	0.546
Na+	0.469
Mg2+	0.0528
SO2–4	0.0282
Ca2+	0.0103
K+	0.0102
C$_T$	0.00206
Br–	0.000844
B$_T$	0.000416
Sr2+	0.000091
F–	0.000068

Scientific theories behind the origins of sea salt started with Sir Edmond Halley in 1715, who proposed that salt and other minerals were carried into the sea by rivers after rainfall washed it out of the ground. Upon reaching the ocean, these salts concentrated as more salt arrived over time. Halley noted that most lakes that don't have ocean outlets (such as the Dead Sea and the Caspian Sea, endorheic basin), have high salt content. Halley termed this process "continental weathering".

Halley's theory was partly correct. In addition, sodium leached out of the ocean floor when the ocean formed. The presence of salt's other dominant ion, chloride, results from outgassing of chloride (as hydrochloric acid) with other gases from Earth's interior via volcanos and hydrothermal vents. The sodium and chloride ions subsequently became the most abundant constituents of sea salt.

Ocean salinity has been stable for billions of years, most likely as a consequence of a chemical/tectonic system which removes as much salt as is deposited; for instance, sodium and chloride sinks include evaporite deposits, pure water burial, and reactions with seafloor basalts.

Human Impacts

Climate change, rising atmospheric carbon dioxide, excess nutrients, and pollution in many forms are altering global oceanic geochemistry. Rates of change for some aspects greatly exceed those in the historical and recent geological record. Major trends include an increasing acidity, reduced subsurface oxygen in both near-shore and pelagic waters, rising coastal nitrogen levels, and widespread increases in mercury and persistent organic pollutants. Most of these perturbations are tied either directly or indirectly to human fossil fuel combustion, fertilizer, and industrial activity. Concentrations are projected to grow in coming decades, with negative impacts on ocean biota and other marine resources.

Human Consumption

Accidentally consuming small quantities of clean seawater is not harmful, especially if the seawater is taken along with a larger quantity of fresh water. However, drinking seawater to maintain hydration is counterproductive; more water must be excreted to eliminate the salt (via urine) than the amount of water obtained from the seawater itself.

The renal system actively regulates sodium chloride in the blood within a very narrow range around 9 g/L (0.9% by weight).

In most open waters concentrations vary somewhat around typical values of about 3.5%, far higher than the body can tolerate and most beyond what the kidney can process. A point frequently overlooked, in claims that the kidney can excrete NaCl in Baltic concentrations (2%), is that the gut cannot absorb water at such concentrations, so that there is no benefit in drinking such water. Drinking seawater temporarily increases blood's NaCl concentration. This signals the kidney to excrete sodium, but seawater's sodium concentration is above the kidney's maximum concentrating ability. Eventually the blood's sodium concentration rises to toxic levels, removing water from cells and interfering with nerve conduction, ultimately producing fatal seizure and cardiac arrhythmia.

Survival manuals consistently advise against drinking seawater. A summary of 163 life raft voyages estimated the risk of death at 39% for those who drank seawater, compared to 3% for those who did not. The effect of seawater intake on rats confirmed the negative effects of drinking seawater when dehydrated.

The temptation to drink seawater was greatest for sailors who had expended their supply of fresh water, and were unable to capture enough rainwater for drinking. This frustration was described famously by a line from Samuel Taylor Coleridge's *The Rime of the Ancient Mariner*:

> *"Water, water, everywhere,*
>
> *And all the boards did shrink;*
>
> *Water, water, everywhere,*
>
> *Nor any drop to drink."*

Although humans cannot survive on seawater, some people claim that up to two cups a day, mixed with fresh water in a 2:3 ratio, produces no ill effect. The French physician Alain Bombard survived an ocean crossing in a small Zodiak rubber boat using mainly raw fish meat, which contains about 40 percent water (like most living tissues), as well as small amounts of seawater and other provisions harvested from the ocean. His findings were challenged, but an alternative explanation was not given. In his 1948 book, *Kon-Tiki*, Thor Heyerdahl reported drinking seawater mixed with fresh in a 2:3 ratio during the 1947 expedition. A few years later, another adventurer, William Willis, claimed to have drunk two cups of seawater and one cup of fresh per day for 70 days without ill effect when he lost part of his water supply.

During the 18th Century, Richard Russell advocated the practice's medical use in the UK, and René Quinton expanded the advocation of the practice other countries, notably France, in the 20th century. Currently, the practice is widely used in Nicaragua and other countries, supposedly taking advantage of the latest medical discoveries.

Most ocean-going vessels desalinate potable water from seawater using processes such as vacuum distillation or multi-stage flash distillation in an evaporator, or more recently by reverse osmosis. These energy-intensive processes were not usually available during the Age of Sail. Larger sailing warships with large crews, such as Nelson's HMS *Victory*, were fitted with distilling apparati in their galleys. Animals such as fish, whales, sea turtles, and seabirds, such as penguins and albatrosses, can adapt to a high saline habitat. For example, the kidney of the desert rat can concentrate sodium far more efficiently than the human kidney.

Standard

ASTM International has an international standard for artificial seawater: ASTM D1141-98 (Original Standard ASTM D1141-52). It is used in many research testing labs as a reproducible solution for seawater such as tests on corrosion, oil contamination, and detergency evaluation.

Surface Water

Image of the entire surface water flow of the Alapaha River near Jennings, Florida going into a sinkhole leading to the Floridan Aquifer groundwater.

Surface water is water on the surface of the planet such as in a river, lake, wetland, or ocean. It can be contrasted with groundwater and atmospheric water.

Non-saline surface water is replenished by precipitation and by recruitment from ground-water. It is lost through evaporation, seepage into the ground where it becomes ground-water, used by plants for transpiration, extracted by mankind for agriculture, living, industry etc. or discharged to the sea where it becomes saline.

Conjunctive Use of Ground and Surface Water

Surface and groundwater are two separate entities, so they must be regarded as such. However, there is an ever-increasing need for management of the two as they are part of an interrelated system that is paramount when the demand for water exceeds the available supply (Fetter 464). Depletion of surface and ground water sources for public consumption (including industrial, commercial, and residential) is caused by over-pumping. Aquifers near river systems that are over-pumped have been known to deplete surface water sources as well. Research supporting this has been found in numerous water budgets for a multitude of cities.

Response times for an aquifer are long (Young & Bredehoeft 1972). However, a total ban on ground water usage during water recessions would allow surface water to better retain levels required for sustainable aquatic life. By reducing ground water pumping, the surface water supplies will be able to maintain their levels, as they recharge from direct precipitation, surface runoff, etc.

Sub-field of Surface Water

Evapotranspiration

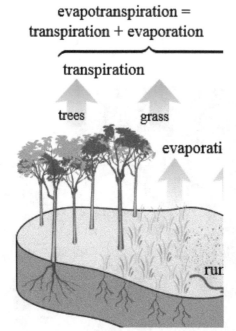

Water cycle of the Earth's surface, showing the individual components of transpiration and evaporation that make up evapotranspiration. Other closely related processes shown are runoff and groundwater recharge.

Evapotranspiration (ET) is the sum of evaporation and plant transpiration from the Earth's land and ocean surface to the atmosphere. Evaporation accounts for the movement of water to the air from sources such as the soil, canopy interception, and waterbodies. Transpiration accounts for the movement of water within a plant and the subsequent loss of water as vapor through stomata in its leaves. Evapotranspiration is an important part of the water cycle. An element (such as a tree) that contributes to evapotranspiration can be called an evapotranspirator.

Reference evapotranspiration (ETo), sometimes incorrectly referred to as potential ET, is a representation of the environmental demand for evapotranspiration and represents the evapotranspiration rate of a short green crop (grass), completely shading the ground, of uniform height and with adequate water status in the soil profile. It is a reflection of the energy available to evaporate water, and of the wind available to transport the water vapour from the ground up into the lower atmosphere. Actual evapotranspiration is said to equal reference evapotranspiration when there is ample water. Some US states utilize a full cover alfalfa reference crop that is 0.5 m in height, rather than the short green grass reference, due to the higher value of ET from the alfalfa reference.

Water Cycle

Evapotranspiration is a significant water loss from drainage basins. Types of vegetation and land use significantly affect evapotranspiration, and therefore the amount of water leaving a drainage basin. Because water transpired through leaves comes from the roots, plants with deep reaching roots can more constantly transpire water. Herbaceous plants generally transpire less than woody plants because they usually have less extensive foliage. Conifer forests tend to have higher rates of evapotranspiration than deciduous forests, particularly in the dormant and early spring seasons. This is primarily due to the enhanced amount of precipitation intercepted and evaporated by conifer foliage during these periods. Factors that affect evapotranspiration include the plant's growth stage or level of maturity, percentage of soil cover, solar radiation, humidity, temperature, and wind. Isotope measurements indicate transpiration is the larger component of evapotranspiration.

Through evapotranspiration, forests reduce water yield, except in unique ecosystems called cloud forests. Trees in cloud forests collect the liquid water in fog or low clouds onto their surface, which drips down to the ground. These trees still contribute to evapotranspiration, but often collect more water than they evaporate or transpire.

In areas that are not irrigated, actual evapotranspiration is usually no greater than precipitation, with some buffer in time depending on the soil's ability to hold water. It will usually be less because some water will be lost due to percolation or surface runoff. An exception is areas with high water tables, where capillary action can cause water from the groundwater to rise through the soil matrix to the surface. If potential evapotranspiration is greater than actual precipitation, then soil will dry out, unless irrigation is used.

Evapotranspiration can never be greater than PET, but can be lower if there is not enough water to be evaporated or plants are unable to transpire readily.

Estimating Evapotranspiration

Evapotranspiration can be measured or estimated using several methods.

Indirect Methods

Pan evaporation data can be used to estimate lake evaporation, but transpiration and evaporation of intercepted rain on vegetation are unknown. There are three general approaches to estimate evapotranspiration indirectly.

Catchment Water Balance

Evapotranspiration may be estimated by creating an equation of the water balance of a drainage basin. The equation balances the change in water stored within the basin (S) with inputs and outgoes:

$$\Delta S = P - ET - Q - D$$

The input is precipitation (P) and the outgoes are evapotranspiration (which is to be estimated), streamflow (Q), and groundwater recharge (D). If the change in storage, precipitation, streamflow, and groundwater recharge are all estimated, the missing flux, ET, can be estimated by rearranging the above equation as follows:

$$ET = P - \Delta S - Q - D$$

Hydrometeorological Equations

The most general and widely used equation for calculating reference ET is the Penman equation. The Penman-Monteith variation is recommended by the Food and Agriculture Organization and the American Society of Civil Engineers. The simpler Blaney-Criddle equation was popular in the Western United States for many years but it is not as accurate in regions with higher humidities. Other solutions used includes Makkink, which is simple but must be calibrated to a specific location, and Hargreaves. To convert the reference evapotranspiration to actual crop evapotranspiration, a crop coefficient and a stress coefficient must be used. Crop coefficients referred to in many hydrological models are themselves during periods for which the model is used. This is because crops are seasonal, perennial plants mature over multiple seasons, and stress responses can significantly depend upon many aspects of plant condition.

Energy Balance

A third methodology to estimate the actual evapotranspiration is the use of the energy balance.

$$\lambda E = R_n - G - H$$

where λE is the energy needed to change the phase of water from liquid to gas, R_n is the net radiation, G is the soil heat flux and H is the sensible heat flux. Using instruments like a scintillometer, soil heat flux plates or radiation meters, the components of the energy balance can be calculated and the energy available for actual evapotranspiration can be solved.

The SEBAL and METRIC algorithms solve the energy balance at the earth's surface using satellite imagery. This allows for both actual and potential evapotranspiration to be calculated on a pixel-by-pixel basis. Evapotranspiration is a key indicator for water management and irrigation performance. SEBAL and METRIC can map these key indicators in time and space, for days, weeks or years.

Experimental Methods for Measuring Evapotranspiration

One method for measuring evapotranspiration is with a weighing lysimeter. The weight of a soil column is measured continuously and the change in storage of water in the soil is modeled by the change in weight. The change in weight is converted to units of length using the surface area of the weighing lysimeter and the unit weight of water. evapotranspiration is computed as the change in weight plus rainfall minus percolation.

Remote Sensing

In recent decades, estimating evapotranspiration has been improved by advances in remote sensing, particularly in agricultural studies. However, quantifying evapotranspiration from mixed vegetation environs, particularly urban parklands, is still challenging because of the heterogeneity of plant species, canopy covers and microclimates and because the methodology is costly. Different remote sensing-based approaches for estimating evapotranspiration have various advantages and disadvantages.

Eddy Covariance

The most direct method of measuring evapotranspiration is with the eddy covariance technique in which fast fluctuations of vertical wind speed are correlated with fast fluctuations in atmospheric water vapor density. This directly estimates the transfer of water vapor (evapotranspiration) from the land (or canopy) surface to the atmosphere.

Urban Landscape Plants

Methods for measuring evapotranspiration can be adapted to an urban setting to estimate the water requirements of urban landscape vegetation.

Potential Evapotranspiration

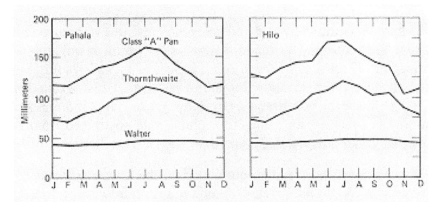

Monthly estimated potential evapotranspiration and measured pan evaporation for two locations in Hawaii, Hilo and Pahala.

Potential evapotranspiration (PET) is the amount of water that would be evaporated and transpired if there were sufficient water available. This demand incorporates the energy available for evaporation and the ability of the lower atmosphere to transport evaporated moisture away from the land surface. Potential evapotranspiration is higher in the summer, on less cloudy days, and closer to the equator, because of the higher levels of solar radiation that provides the energy for evaporation. Potential evapotranspiration is also higher on windy days because the evaporated moisture can be quickly moved from the ground or plant surface, allowing more evaporation to fill its place.

Potential evapotranspiration is expressed in terms of a depth of water, and can be graphed during the year.

Potential evapotranspiration is usually measured indirectly, from other climatic factors, but also depends on the surface type, such as free water (for lakes and oceans), the soil type for bare soil, and the vegetation. Often a value for the potential evapotranspiration is calculated at a nearby climate station on a reference surface, conventionally short grass. This value is called the reference evapotranspiration, and can be converted to a potential evapotranspiration by multiplying with a surface coefficient. In agriculture, this is called a crop coefficient. The difference between potential evapotranspiration and precipitation is used in irrigation scheduling.

Average annual potential evapotranspiration is often compared to average annual precipitation, P. The ratio of the two, P/PET, is the aridity index.

Groundwater Recharge

Groundwater recharge or deep drainage or deep percolation is a hydrologic process where water moves downward from surface water to groundwater. Recharge is the primary method through which water enters an aquifer. This process usually occurs in the vadose zone below plant roots and is often expressed as a flux to the water table surface. Recharge occurs both naturally (through the water cycle) and through anthropogenic processes (i.e., "artificial groundwater recharge"), where rainwater and or reclaimed water is routed to the subsurface.

Processes

Groundwater is recharged naturally by rain and snow melt and to a smaller extent by surface water (rivers and lakes). Recharge may be impeded somewhat by human activities including paving, development, or logging. These activities can result in loss of topsoil resulting in reduced water infiltration, enhanced surface runoff and reduction in recharge. Use of groundwaters, especially for irrigation, may also lower the water tables. Groundwater recharge is an important process for sustainable groundwater management, since the volume-rate abstracted from an aquifer in the long term should be less than or equal to the volume-rate that is recharged.

Recharge can help move excess salts that accumulate in the root zone to deeper soil layers, or into the groundwater system. Tree roots increase water saturation into groundwater reducing water runoff. Flooding temporarily increases river bed permeability by moving clay soils downstream, and this increases aquifer recharge.

Artificial groundwater recharge is becoming increasingly important in India, where over-pumping of groundwater by farmers has led to underground resources becoming depleted. In 2007, on the recommendations of the International Water Management Institute, the Indian government allocated Rs 1800 crore (US$400million) to fund dug-well recharge projects (a dug-well is a wide, shallow well, often lined with concrete) in 100 districts within seven states where water stored in hard-rock aquifers had been over-exploited. Another environmental issue is the disposal of waste through the water flux such as dairy farms, industrial, and urban runoff.

Wetlands

Wetlands help maintain the level of the water table and exert control on the hydraulic head (O'Brien 1988; Winter 1988). This provides force for groundwater recharge and discharge to other waters as well. The extent of groundwater recharge by a wetland is dependent upon soil, vegetation, site, perimeter to volume ratio, and water table gradient (Carter and Novitzki 1988; Weller 1981). Groundwater recharge occurs through mineral soils found primarily around the edges of wetlands (Verry and Timmons 1982) The soil under most wetlands is relatively impermeable. A high perimeter to volume ratio, such as in small wetlands, means that the surface area through which water can infiltrate into the groundwater is high (Weller 1981). Groundwater recharge is typical in small wetlands such as prairie potholes, which can contribute significantly to recharge of regional groundwater resources (Weller 1981). Researchers have discovered groundwater recharge of up to 20% of wetland volume per season (Weller 1981).

Estimation Methods

Rates of groundwater recharge are difficult to quantify since other related processes, such as evaporation, transpiration (or evapotranspiration) and infiltration processes must first be measured or estimated to determine the balance.

Physical

Physical methods use the principles of soil physics to estimate recharge. The *direct* physical methods are those that attempt to actually measure the volume of water passing below the root zone. *Indirect* physical methods rely on the measurement or estimation of soil physical parameters, which along with soil physical principles, can be used to estimate the potential or actual recharge. After months without rain the level of the rivers under humid climate is low and represents solely drained groundwater. Thus, the recharge can be calculated from this base flow if the catchment area is known.

Chemical

Chemical methods use the presence of relatively inert water-soluble substances, such as an isotopic tracer or chloride, moving through the soil, as deep drainage occurs.

Numerical Models

Recharge can be estimated using numerical methods, using such codes as Hydrologic Evaluation of Landfill Performance, UNSAT-H, SHAW, WEAP, and MIKE SHE. The 1D-program HYDRUS1D is available online. The codes generally use climate and soil data to arrive at a recharge estimate and use the Richards equation in some form to model groundwater flow in the vadose zone.

References

- National Research Council Staff (1995). Mexico City's Water Supply: Improving the Outlook for Sustainability. Washington, D.C., USA: National Academies Press. ISBN 978-0-309-05245-0.

- G. Mokhtar (1981-01-01). Ancient civilizations of Africa. Books.google.com. Unesco. International Scientific Committee for the Drafting of a General History of Africa. p. 309. ISBN 9780435948054. Retrieved 2012-06-19.

- Richard Bulliet, Pamela Kyle Crossley, Daniel Headrick, Steven Hirsch. Pages 53-56 (2008-06-18). The Earth and Its Peoples, Volume I: A Global History, to 1550. Books.google.com. ISBN 0618992383. Retrieved 2012- 06-19.

- Frenken, K. (2005). Irrigation in Africa in figures – AQUASTAT Survey – 2005 (PDF). Food and Agriculture Organization of the United Nations. ISBN 92-5-105414-2. Retrieved 2007-03-14.

- Drainage Manual: A Guide to Integrating Plant, Soil, and Water Relationships for Drainage of Irrigated Lands. Interior Dept., Bureau of Reclamation. 1993. ISBN 0-16-061623-9.

- Stumm, W, Morgan, J. J. (1981) Aquatic Chemistry, An Introduction Emphasizing Chemical Equilibria in Natural Waters. John Wiley & Sons. pp. 414–416. ISBN 0471048313.

- Rippon, P.M., Commander, RN (1998). The evolution of engineering in the Royal Navy. Vol 1: 1827–1939. Spellmount. pp. 78–79. ISBN 0-946771-55-3.

- Sneed, M; Brandt, J; Solt, M (2013). "Land Subsidence along the Delta-Mendota Canal in the Northern Part of the San Joaquin Valley, California, 2003–10" (PDF). USGS Scientific Investigations Report 2013-5142. Retrieved 22 June 2015.

- "What is hydrology and what do hydrologists do?". The USGS Water Science School. United States Geological Survey. 23 May 2013. Retrieved 21 Jan 2014.

- Tosi, Luigi; Teatini, Pietro; Strozzi, Tazio; Da Lio, Cristina (2014). "Relative Land Subsidence of the Venice Coastland, Italy": 171–173. doi:10.1007/978-3-319-08660-6_32.

- "Africa, Emerging Civilizations In Sub-Sahara Africa. Various Authors; Edited By: R. A. Guisepi". History-world.org. Retrieved 2012-06-19.

- Mader, Shelli (May 25, 2010). "Center pivot irrigation revolutionizes agriculture". The Fence Post Magazine. Retrieved June 6, 2012.

- "Urban Trees Enhance Water Infiltration". Fisher, Madeline. The American Society of Agronomy. November 17, 2008. Retrieved October 31, 2012.

3

Study of Fresh Water

The chapter illustrates and demonstrates the study of fresh water. Fresh water is naturally occurring water on earth's surface in ice sheets, ice caps, rivers and streams, and groundwater. The following content also gives a profound content on desalination. The topics discussed in the chapter are of great importance to broaden the existing knowledge on water resources.

Fresh water is naturally occurring water on Earth's surface in ice sheets, ice caps, glaciers, icebergs, bogs, ponds, lakes, rivers and streams, and underground as groundwater in aquifers and underground streams. Fresh water is generally characterized by having low concentrations of dissolved salts and other total dissolved solids. The term specifically excludes seawater and brackish water although it does include mineral-rich waters such as chalybeate springs. The term "sweet water" (from Spanish "agua dulce") has been used to describe fresh water in contrast to salt water. The term fresh water does not have the same meaning as potable water. Much of the surface fresh water and ground water is unsuitable for drinking without some form of purification because of the presence of chemical or biological contaminants.

Earth seen from Apollo 17 — the Antarctic ice sheet at the bottom of the photograph contains 61% of the fresh water, or 1.7% of the total water, on Earth.

Systems

Scientifically, fresh water habitats are divided into lentic systems, which are the stillwaters including ponds, lakes, swamps and mires; lotic systems, which are running water; and groundwater which flows in rocks and aquifers. There is, in addition, a zone which bridges between groundwater and lotic systems, which is the hyporheic zone, which underlies many larger rivers and can contain substantially more water than is seen in the open channel. It may also be in direct contact with the underlying underground water. The majority of fresh water is in icecaps.

The surface of a fresh water lake.

Sources

The source of almost all fresh water is precipitation from the atmosphere, in the form of mist, rain and snow. Fresh water falling as mist, rain or snow contains materials dissolved from the atmosphere and material from the sea and land over which the rain bearing clouds have traveled. In industrialized areas rain is typically acidic because of dissolved oxides of sulfur and nitrogen formed from burning of fossil fuels in cars, factories, trains and aircraft and from the atmospheric emissions of industry. In some cases this acid rain results in pollution of lakes and rivers.

In coastal areas fresh water may contain significant concentrations of salts derived from the sea if windy conditions have lifted drops of seawater into the rain-bearing clouds. This can give rise to elevated concentrations of sodium, chloride, magnesium and sulfate as well as many other compounds in smaller concentrations.

In desert areas, or areas with impoverished or dusty soils, rain-bearing winds can pick up sand and dust and this can be deposited elsewhere in precipitation and causing the freshwater flow to be measurably contaminated both by insoluble solids but also by the soluble components of those soils. Significant quantities of iron may be transported in this way including the well-documented transfer of iron-rich rainfall falling in Brazil derived from sand-storms in the Sahara in north Africa.

Water Distribution

Water is a critical issue for the survival of all living organisms. Some can use salt water but many organisms including the great majority of higher plants and most mammals must have access to fresh water to live. Some terrestrial mammals, especially desert rodents appear to survive without drinking but they do generate water through the metabolism of cereal seeds and they also have mechanisms to conserve water to the maximum degree.

Out of all the water on Earth, saline water in oceans, seas and saline groundwater make up about 97% of it. Only 2.5–2.75% is fresh water, including 1.75–2% frozen in glaciers, ice and snow, 0.5–0.75% as fresh groundwater and soil moisture, and less than 0.01% of it as surface water in lakes, swamps and rivers. Freshwater lakes contain about 87% of this fresh surface water, including 29%

in the African Great Lakes, 20% in Lake Baikal in Russia, 21% in the North American Great Lakes, and 14% in other lakes. Swamps have most of the balance with only a small amount in rivers, most notably the Amazon River. The atmosphere contains 0.04% water. In areas with no fresh water on the ground surface, fresh water derived from precipitation may, because of its lower density, over-lie saline ground water in lenses or layers. Most of the world's fresh water is frozen in ice sheets. Many areas suffer from lack of distribution of fresh water, such as deserts.

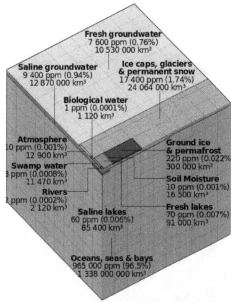

Visualisation of the distribution (by volume) of water on Earth. Each tiny cube (such as the one representing biological water) corresponds to approximately 1000 cubic km of water, with a mass of approximately 1 trillion tonnes (200000 times that of the Great Pyramid of Giza or 5 times that of Lake Kariba, arguably the heaviest man-made object). The entire block comprises 1 million tiny cubes.

Numerical Definition

Fresh water can be defined as water with less than 500 parts per million (ppm) of dissolved salts.

Water salinity based on dissolved salts			
Fresh water	**Brackish water**	**Saline water**	**Brine**
< 0.05%	0.05–3%	3–5%	> 5%

Other sources give higher upper salinity limits for fresh water, e.g. 1000 ppm or 3000 ppm.

Aquatic Organisms

Fresh water creates a hypotonic environment for aquatic organisms. This is problematic for some organisms with pervious skins or with gill membranes, whose cell membranes may burst if excess water is not excreted. Some protists accomplish this using contractile vacuoles, while freshwater fish excrete excess water via the kidney. Although most aquatic organisms have a limited ability to regulate their osmotic balance and therefore can only live within a narrow range of salinity, dia-dromous fish have the ability to migrate between fresh water and saline water bodies. During these migrations they undergo changes to adapt to the surroundings of the changed salinities; these processes are hormonally controlled. The eel (*Anguilla anguilla*) uses the hormone prolactin, while in salmon (*Salmo salar*) the hormone cortisol plays a key role during this process.

Many sea birds have special glands at the base of the bill through which excess salt is excreted. Similarly the marine iguanas on the Galápagos Islands excrete excess salt through a nasal gland and they sneeze out a very salty excretion.

Fresh Water as a Resource

Water fountain found in a small Swiss village. They are used as a drinking water source for people and cattle. Almost every Alpine village has such a water source.

An important concern for hydrological ecosystems is securing minimum streamflow, especially preserving and restoring instream water allocations. Fresh water is an important natural resource necessary for the survival of all ecosystems. The use of water by humans for activities such as irrigation and industrial applications can have adverse impacts on down-stream ecosystems. Chemical contamination of fresh water can also seriously damage eco-systems.

Pollution from human activity, including oil spills and also presents a problem for freshwater resources. The largest petroleum spill that has ever occurred in fresh water was caused by a Royal Dutch Shell tank ship in Magdalena, Argentina, on 15 January 1999, polluting the environment, drinkable water, plants and animals.

Fresh and unpolluted water accounts for 0.003% of total water available globally.

Agriculture

Changing landscape for the use of agriculture has a great effect on the flow of fresh water. Changes in landscape by the removal of trees and soils changes the flow of fresh water in the local environment and also affects the cycle of fresh water. As a result, more fresh water is stored in the soil which benefits agriculture. However, since agriculture is the human activity that consumes the most fresh water, this can put a severe strain on local freshwater resources resulting in the destruction of local ecosystems. In Australia, over-abstraction of fresh water for intensive irrigation activities has caused 33% of the land area to be at risk of salination. With regards to agriculture, the World Bank targets food production and water management as an increasingly global issue that will foster debate.

Limited Resource

Fresh water is a renewable and variable, but finite natural resource. Fresh water can only be replenished through the process of the water cycle, in which water from seas, lakes, forests, land, rivers, and reservoirs evaporates, forms clouds, and returns as precipitation. Locally however, if more fresh water is consumed through human activities than is naturally restored, this may result in reduced fresh water availability from surface and underground sources and can cause serious damage to surrounding and associated environments.

Fresh Water Withdrawal

Fresh water withdrawal is the quantity of water removed from available sources for use in any purpose, excluding evaporation losses. Water drawn off is not necessarily entirely consumed and some portion may be returned for further use downstream.

Causes of Limited Fresh Water

There are many causes of the apparent decrease in our fresh water supply. Principal amongst these is the increase in population through increasing life expectancy, the increase in per capita water use and the desire of many people to live in warm climates that have naturally low levels of fresh water resources. Climate change is also likely to change the availability and distribution of fresh water across the planet:

"If global warming continues to melt glaciers in the polar regions, as expected, the supply of fresh water may actually decrease. First, fresh water from the melting glaciers will mingle with salt water in the oceans and become too salty to drink. Second, the increased ocean volume will cause sea levels to rise, contaminating freshwater sources along coastal regions with seawater".

The World Bank adds that the response by freshwater ecosystems to a changing climate can be described in terms of three interrelated components: water quality, water quantity or volume, and water timing. A change in one often leads to shifts in the others as well. Water pollution and subsequent eutrophication also reduces the availability of fresh water.

Also, there is an uneven distribution of fresh water. While some countries have an abundant supply of fresh water, others do not have as much. For example, Canada has 20% of the world's fresh water supply, while India has only 10% of the world's fresh water supply, even though India's population is more than 30 times larger than that of Canada. A reason for the uneven distribution of fresh water supply may be the differences in climate. For example, in some countries in Africa, the frequent lack of rain has led to insufficient water supply for irrigation. This has affected agriculture and has led to a shortage of food for the people.

Fresh Water in the Future

Many areas of the world are already experiencing stress on water availability. Due to the accelerated pace of population growth and an increase in the amount of water a single person uses, it is expected that this situation will continue to get worse. A shortage of water in the future would be detrimental to the human population as it would affect everything from sanitation, to overall health and the production of grain.

Choices in the use of Fresh Water

With one in eight people in the world not having access to safe water it is important to use this resource in a prudent manner. Making the best use of water on a local basis probably provides the best solution. Local communities need to plan their use of fresh water and should be made aware of how certain crops and animals use water.

As a guide the following tables provide some indicators.

Table 1. Recommended basic water requirements for human needs (per person)

Activity	Minimum, litres / day	Range / day
Drinking Water	5	2–5
Sanitation Services	20	20–75
Bathing	15	5–70
Cooking and Kitchen	10	10–50

Table 2. Water Requirements of different classes of livestock

Animal	Average / day	Range / day
Dairy cow	76 L (20 US gal)	57 to 95 L (15 to 25 US gal)
Cow-calf pair	57 L (15 US gal)	8 to 76 L (2 to 20 US gal)
Yearling cattle	38 L (10 US gal)	23 to 53 L (6 to 14 US gal)
Horse	38 L (10 US gal)	30 to 53 L (8 to 14 US gal)
Sheep	8 L (2 US gal)	8 to 11 L (2 to 3 US gal)

Table 3. Approximate values of seasonal crop water needs

Crop	Crop water needs mm / total growing period
Banana	1200–2200
Barley/Oats/Wheat	450–650
Cabbage	350–500
Citrus	900–1200

Onions	350–550
Pea	350–500
Potato	500–700
Sugar Cane	1500–2500
Tomato	400–800

Accessing Fresh Water

Canada

Canada has approximately 7% of the world's renewable fresh water. Canadians access their water from ground water, lakes and streams; it is then cleaned and purified in water treatment plants.

United States

The United States uses much more water per capita than developing countries. For example, the average American's daily shower uses more water than a person in a developing country would use for an entire day. Las Vegas, a city that uses an extreme amount of water to support spectacular lush greenery and golf courses, as well as huge fountains and swimming pools gets 90% of its water from Lake Mead, which is now at a record all-time low.

Developing Countries

In developing countries, 780 million people lack access to clean water. Half of the population of the developing world suffers from at least one disease caused by insufficient water supply and sanitation.

Sub-field Of Fresh Water

Desalination

Water Desalination

Methods

- Distillation
    - Multi-stage flash distillation (MSF)
    - Multiple-effect distillation (MED | ME)
    - Vapor-compression (VC)
- Ion exchange
- Membrane processes
    - Electrodialysis reversal (EDR)
    - Reverse osmosis (RO)
    - Nanofiltration (NF)

- o   Membrane distillation (MD)

- o   Forward osmosis (FO)

- Freezing desalination

- Geothermal desalination

- Solar desalination

    - o   Solar humidification-Dehumidification (HDH)

    - o   Multiple-effect humidification (MEH)

- Methane hydrate crystallization

- High grade water recycling

- Seawater greenhouse Desalination is a process that removes minerals from saline water. More generally, desalination refers to the removal of salts and minerals from a target substance, as in soil desalination, which is an issue for agriculture.

Saltwater is desalinated to produce water suitable for human consumption or irrigation. One by-product of desalination is salt. Desalination is used on many seagoing ships and submarines. Most of the modern interest in desalination is focused on cost-effective provision of fresh water for human use. Along with recycled wastewater, it is one of the few rainfall-independent water sources.

Due to its energy consumption, desalinating sea water is generally more costly than fresh water from rivers or groundwater, water recycling and water conservation. However, these alternatives are not always available and depletion of reserves is a critical problem worldwide. Currently, approximately 1% of the world's population is dependent on desalinated water to meet daily needs, but the UN expects that 14% of the world's population will encounter water scarcity by 2025.

Desalination is particularly relevant in dry countries such as Australia, which traditionally have relied on collecting rainfall behind dams for water.

According to the International Desalination Association, in June 2015, 18,426 desalination plants operated worldwide, producing 86.8 million cubic meters per day, providing water for 300 million people. This number increased from 78.4 million cubic meters in 2013, a 57% increase in just 5 years. The single largest desalination project is Ras Al-Khair in Saudi Arabia, which produced 1,025,000 cubic meters per day in 2014, although this plant is expected to be surpassed by a plant in California. Israel produces a higher proportion of its water than any other country, totaling 40% of its water use.

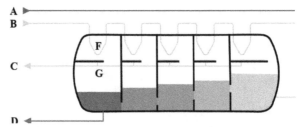

Schematic of a multistage flash desalinatorA – steam inB – seawater inC – potable water outD – waste outE – steam outF – heat exchangeG – condensation collectionH – brine heater

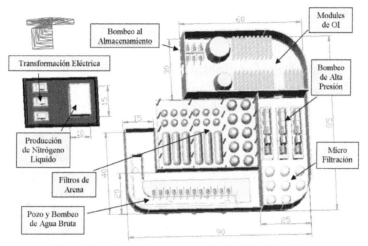

Plan of a typical reverse osmosis desalination plant

Methods

The traditional process used in these operations is vacuum distillation—essentially boiling it to leave impurities behind. In desalination, atmospheric pressure is reduced, thus lowering the required temperature. Liquids boil when the vapor pressure equals the ambient pressure and vapor pressure increases with temperature. Thus, because of the reduced temperature, low-temperature "waste" heat from electrical power generation or industrial processes can be employed.

Reverse osmosis desalination plant in Barcelona, Spain

The principal competing processes use membranes to desalinate, principally applying reverse osmosis. Membrane processes use semipermeable membranes and pressure to separate salts from water. Reverse osmosis plant membrane systems typically use less energy than thermal distillation. Desalination remains energy intensive, however, and future costs will continue to depend on the energy prices.

Considerations and Criticism

Energy Consumption

Energy consumption of seawater desalination has reached as low as 3 kWh/m³, including pre-filtering and ancillaries, similar to the energy consumption of other fresh water supplies transported over large distances, but much higher than local fresh water supplies that use 0.2 kWh/m³ or less.

A minimum energy consumption for seawater desalination of around 1 kWh/m³ has been determined, excluding prefiltering and intake/outfall pumping. Under 2 kWh/m³ has been achieved with reverse osmosis membrane technology, leaving limited scope for further energy reductions.

Supplying all US domestic water by desalination would increase energy consumption by around 10%, about the amount of energy used by domestic refrigerators. Domestic consumption is a relatively small fraction of the total water usage.

Energy consumption of seawater desalination methods.				
Desalination Method >>	**Multi-stage Flash MSF**	**Multi-Effect Distillation MED**	**Mechanical Vapor Compression MVC**	**Reverse Osmosis RO**
Electrical energy (kWh/m³)	4–6	1.5–2.5	7–12	3–5.5
Thermal energy (kWh/m³)	50–110	60–110	None	None
Electrical equivalent of thermal energy (kWh/m³)	9.5–19.5	5–8.5	None	None
Total equivalent electrical energy (kWh/m³)	13.5–25.5	6.5–11	7–12	3–5.5

Note: "Electrical equivalent" refers to the amount of electrical energy that could be generated using a given quantity of thermal energy and appropriate turbine generator. These calculations do not include the energy required to construct or refurbish items consumed in the process.

Cogeneration

Cogeneration is generating excess heat and electricity generation from a single process. Cogeneration can provide usable heat for desalination in an integrated, or "dual-purpose", facility where a power plant provides the energy for desalination. Alternatively, the facility's energy production may be dedicated to the production of potable water (a stand-alone facility), or excess energy may be produced and incorporated into the energy grid. Cogeneration takes various forms, and theoretically any form of energy production could be used. However, the majority of current and planned cogeneration desalination plants use either fossil fuels or nuclear power as their source of energy. Most plants are located in the Middle East or North Africa, which use their petroleum resources to offset limited water resources. The advantage of dual-purpose facilities is they can be more efficient in energy consumption, thus making desalination more viable.

The current trend in dual-purpose facilities is hybrid configurations, in which the permeate from reverse osmosis desalination is mixed with distillate from thermal desalination. Basically, two or more desalination processes are combined along with power production. Such facilities have been implemented in Saudi Arabia at Jeddah and Yanbu.

The Shevchenko BN350, a nuclear-heated desalination unit

A typical Supercarrier in the US military uses nuclear power to desalinate 400,000 US gallons (1,500,000 l; 330,000 imp gal) of water per day.

Economics

Costs of desalinating sea water (infrastructure, energy, and maintenance) are generally higher than fresh water from rivers or groundwater, water recycling, and water conservation, but alternatives are not always available. Desalination costs in 2013 ranged from US$0.45 to $1.00/cubic metre ($US2 to 4/kgal). (1 cubic meter is about 264 gallons.) More than half of the cost comes directly from energy cost, and since energy prices are very volatile, actual costs can vary substantially.

The cost of untreated fresh water in the developing world can reach US$5/cubic metre.

Average water consumption and cost of supply by sea water desalination at US$1 per cubic metre(±50%)		
Area	Consumption USgal/person/day	Consumption litre/person/day	Desalinated Water Cost US$/person/day
USA	100	378	0.38
Europe	50	189	0.19
Africa	15	57	0.06
UN recommended minimum	13	49	0.05

Factors that determine the costs for desalination include capacity and type of facility, location, feed water, labor, energy, financing and concentrate disposal. Desalination stills control pressure, temperature and brine concentrations to optimize efficiency. Nuclear-powered desalination might be economical on a large scale.

While noting costs are falling, and generally positive about the technology for affluent areas in proximity to oceans, a 2004 study argued, "Desalinated water may be a solution for some water-stress regions, but not for places that are poor, deep in the interior of a continent, or at high elevation. Unfortunately, that includes some of the places with biggest water problems.", and, "Indeed, one needs to lift the water by 2,000 m (6,600 ft), or transport it over more than 1,600 km (990 mi) to get transport costs equal to the desalination costs. Thus, it may be more economical to transport fresh water from somewhere else than to desalinate it. In places far from the sea, like New Delhi, or in high places, like Mexico City, transport costs could match desalination costs. Desalinated water is also expensive in places that are both somewhat far from the sea and somewhat high, such as Riyadh and Harare. By contrast in other locations transport costs are much less, such as Beijing, Bangkok, Zaragoza, Phoenix, and, of course, coastal cities like Tripoli." After desalination at Jubail, Saudi Arabia, water is pumped 200 mi (320 km) inland to Riyadh. For coastal cities, desalination is increasingly viewed as a competitive choice.

In 2014, the Israeli facilities of Hadera, Palmahim, Ashkelon, and Sorek were desalinizing water for less than US$0.40 per cubic meter. As of 2006, Singapore was desalinating water for US$0.49 per cubic meter. The city of Perth began operating a reverse osmosis seawater desalination plant in 2006. A desalination plant now operates in Sydney, and the Wonthaggi desalination plant was under construction in Wonthaggi, Victoria.

The Perth desalination plant is powered partially by renewable energy from the Emu Downs Wind Farm. A wind farm at Bungendore in New South Wales was purpose-built to generate enough renewable energy to offset the Sydney plant's energy use, mitigating concerns about harmful greenhouse gas emissions.

In December 2007, the South Australian government announced it would build a seawater desalination plant for the city of Adelaide, Australia, located at Port Stanvac. The desalination plant was to be funded by raising water rates to achieve full cost recovery.

A January 17, 2008, article in the *Wall Street Journal* stated, "In November, Connecticut-based Poseidon Resources Corp. won a key regulatory approval to build the $300 million water-desalination plant in Carlsbad, north of San Diego. The facility would produce 50,000,000 US gallons (190,000,000 l; 42,000,000 imp gal) of drinking water per day, enough to supply about 100,000 homes. As of June 2012, the cost for the desalinated water had risen to $2,329 per acre-foot. Each $1,000 per acre-foot works out to $3.06 for 1,000 gallons, or $.81 per cubic meter.

Poseidon Resources made an unsuccessful attempt to construct a desalination plant in Tampa Bay, FL, in 2001. The board of directors of Tampa Bay Water was forced to buy the plant from Poseidon in 2001 to prevent a third failure of the project. Tampa Bay Water faced five years of engineering problems and operation at 20% capacity to protect marine life. The facility reached capacity only in 2007.

In 2008, a Energy Recovery Inc. was desalinating water for $0.46 per cubic meter.

Environmental

Intake

In the United States, cooling water intake structures are regulated by the Environmental Protection Agency (EPA). These structures can have the same impacts to the environment as desalination facility intakes. According to EPA, water intake structures cause adverse environmental impact by sucking fish and shellfish or their eggs into an industrial system. There, the organisms may be killed or injured by heat, physical stress, or chemicals. Larger organisms may be killed or injured when they become trapped against screens at the front of an intake structure. Alternative intake types that mitigate these impacts include beach wells, but they require more energy and higher costs.

The Kwinana Desalination Plant opened in Perth in 2007. Water there and at Queensland's Gold Coast Desalination Plant and Sydney's Kurnell Desalination Plant is withdrawn at 0.1 m/s (0.33 ft/s), which is slow enough to let fish escape. The plant provides nearly 140,000 m³ (4,900,000 cu ft) of clean water per day.

Outflow

Desalination processes produce large quantities of brine, possibly at above ambient temperature, and contain residues of pretreatment and cleaning chemicals, their reaction byproducts and heavy metals due to corrosion. Chemical pretreatment and cleaning are a necessity in most desalination plants, which typically includes prevention of biofouling, scaling, foaming and corrosion in thermal plants, and of biofouling, suspended solids and scale deposits in membrane plants.

To limit the environmental impact of returning the brine to the ocean, it can be diluted with another stream of water entering the ocean, such as the outfall of a wastewater treatment or power plant. With medium to large power plant and desalination plants, the power plant's cooling water flow is likely to be several times larger than that of the desalination plant, reducing the salinity of the combination. Another method to dilute the brine is to mix it via a diffuser in a mixing zone. For example, once a pipeline containing the brine reaches the sea floor, it can split into many branches, each releasing brine gradually through small holes along its length. Mixing can be combined with power plant or wastewater plant dilution.

Brine is denser than seawater and therefore sinks to the ocean bottom and can damage the ecosystem. Careful reintroduction can minimize this problem. Typical ocean conditions allow for rapid dilution, thereby minimizing harm.

Alternatives without Pollution

Some methods of desalination, particularly in combination with evaporation ponds, solar stills, and condensation trap (solar desalination), do not discharge brine. They do not use chemicals or burn fossil fuels. They do not work with membranes or other critical parts, such as components that include heavy metals, thus do not produce toxic waste (and high maintenance).

A new approach that works like a solar still, but on the scale of industrial evaporation ponds is the integrated biotectural system. It can be considered "full desalination" because it converts the entire amount of saltwater intake into distilled water. One of the advantages of this system is the feasibility for inland operation. Standard advantages also include no air pollution and no temperature increase of endangered natural water bodies from power plant cooling-water discharge. Another important advantage is the production of sea salt for industrial and other uses. As of 2015, 50% of the world's sea salt production relies on fossil energy sources.

Alternatives to Desalination

Increased water conservation and efficiency remain the most cost-effective approaches in areas with a large potential to improve the efficiency of water use practices. Wastewater reclamation provides multiple benefits over desalination. Urban runoff and storm water capture also provide benefits in treating, restoring and recharging groundwater.

A proposed alternative to desalination in the American Southwest is the commercial importation of bulk water from water-rich areas either by oil tankers converted to water carriers, or pipelines. The idea is politically unpopular in Canada, where governments imposed trade barriers to bulk water exports as a result of a North American Free Trade Agreement (NAFTA) claim.

Public Health Concerns

Desalination removes iodine from water and could increase the risk of iodine deficiency disorders. Israeli researchers claimed a possible link between seawater desalination and iodine deficiency, finding deficits among euthyroid adults exposed to iodine-poor water concurrently with an increasing proportion of their area's drinking water from seawater reverse osmosis (SWRO). They later found probable iodine deficiency disorders in a population reliant on desalinated seawater.

Experimental Techniques

Other desalination techniques include:

Waste Heat

Diesel generators commonly provide electricity in remote areas. About 40%–50% of the energy output is low-grade heat that leaves the engine via the exhaust. Connecting a membrane distillation system to the diesel engine exhaust repurposes this low-grade heat for desalination. The system actively cools the diesel generator, improving its efficiency and increasing its electricity output. This results in an energy-neutral desalination solution. An example plant was commissioned by Dutch company Aquaver in March 2014 for Gulhi, Maldives.

Low-temperature Thermal

Originally stemming from ocean thermal energy conversion research, low-temperature thermal desalination (LTTD) takes advantage of water boiling at low pressure, even at ambient temperature. The system uses pumps to create a low-pressure, low-temperature environment in which water boils at a temperature gradient of 8–10 °C (46–50 °F) between two volumes of water. Cool ocean water is supplied from depths of up to 600 m (2,000 ft). This water is pumped through coils to condense the water vapor. The resulting condensate is purified water. LTTD may take advantage of the temperature gradient available at power plants, where large quantities of warm wastewater are discharged from the plant, reducing the energy input needed to create a temperature gradient.

Experiments were conducted in the US and Japan to test the approach. In Japan, a spray-flash evaporation system was tested by Saga University. In Hawaii, the National Energy Laboratory tested an open-cycle OTEC plant with fresh water and power production using a temperature difference of 20 C° between surface water and water at a depth of around 500 m (1,600 ft). LTTD was studied by India's National Institute of Ocean Technology (NIOT) in 2004. Their first LTTD plant opened in 2005 at Kavaratti in the Lakshadweep islands. The plant's capacity is 100,000 L (22,000 imp gal; 26,000 US gal)/day, at a capital cost of INR 50 million (€922,000). The plant uses deep water at a temperature of 10 to 12 °C (50 to 54 °F). In 2007, NIOT opened an experimental, floating LTTD plant off the coast of Chennai, with a capacity of 1,000,000 L (220,000 imp gal; 260,000 US gal)/day. A smaller plant was established in 2009 at the North Chennai Thermal Power Station to prove the LTTD application where power plant cooling water is available.

Thermoionic Process

In October 2009, Saltworks Technologies announced a process that uses solar or other thermal heat to drive an ionic current that removes all sodium and chlorine ions from the water using ion-exchange membranes.

Evaporation and Condensation for Crops

The Seawater greenhouse uses natural evaporation and condensation processes inside a greenhouse powered by solar energy to grow crops in arid coastal land.

Other Approaches

Forward Osmosis

One process was commercialized by Modern Water PLC using forward osmosis, with a number of plants reported to be in operation.

Small-scale Solar

The United States, France and the United Arab Emirates are working to develop practical solar desalination. AquaDania's WaterStillar has been installed at Dahab, Egypt, and in Playa del Carmen, Mexico. In this approach, a solar thermal collector measuring two square metres can distill from 40 to 60 litres per day from any local water source – five times more than conventional stills. It eliminates the need for plastic PET bottles or energy-consuming water transport. In Central California, a startup company WaterFX is developing a solar-powered method of desalination that can enable the use of local water, including runoff water that can be treated and used again. Salty groundwater in the region would be treated to become freshwater, and in areas near the ocean, seawater could be treated.

Passarell

The Passarell process uses reduced atmospheric pressure rather than heat to drive evaporative desalination. The pure water vapor generated by distillation is then compressed and condensed using an advanced compressor. The compression process improves distillation efficiency by creating the reduced pressure in the evaporation chamber. The compressor centrifuges the pure water vapor after it is drawn through a demister (removing residual impurities) causing it to compress against tubes in the collection chamber. The compression of the vapor increases its temperature. The heat is transferred to the input water falling in the tubes, vaporizing the water in the tubes. Water vapor condenses on the outside of the tubes as product water. By combining several physical processes, Passarell enables most of the system's energy to be recycled through its evaporation, demisting, vapor compression, condensation, and water movement processes.

Geothermal

Geothermal energy can drive desalination. In most locations, geothermal desalination beats using scarce groundwater or surface water, environmentally and economically.

Nanotechnology

Nanotube membranes may prove to be effective for water filtration and desalination processes that would require substantially less energy than reverse osmosis.

Hermetic, sulphonated nano-composite membranes have shown to be capable of reducing almost all forms of contamination to the parts per billion level. These nano-materials, using a non-reverse osmosis process, have little or no susceptibility to high salt concentration levels.

Abstracted animation of the nanoscale graphene membrane desalination process.

Biomimesis

Biomimetic membranes are another approach.

Electrochemical

In 2008, Siemens Water Technologies announced technology that applied electric fields to desalinate one cubic meter of water while using only a purported 1.5 kWh of energy. If accurate, this process would consume one-half the energy of other processes. As of 2012 a demonstration plant was operating in Singapore. Researchers at the University of Texas at Austin and the University of Marburg are developing more efficient methods of electrochemically mediated seawater desalination.

Freeze-thaw

Freeze-thaw desalination uses freezing to remove fresh water from frozen seawater.

Electrokinetic Shocks

Membraneless desalination at ambient temperature and pressure used electrokinetic shocks waves. Anions and cations in salt water are exchanged for carbonate anions and calcium cations respectively using electrokinetic shockwaves. Calcium and carbonate ions react to form calcium carbonate, which precipitates, leaving fresh water. Theoretical energy efficiency of this method is on par with electrodialysis and reverse osmosis.

Facilities

Estimates vary widely between 15,000–20,000 desalination plants producing more than 20,000 m³/day. Micro desalination plants operate near almost every natural gas or fracking facility is found in the United States.

In Nature

Mangrove leaf with salt crystals

Evaporation of water over the oceans in the water cycle is a natural desalination process.

The formation of sea ice produces ice with little salt, much lower than in seawater.

Seabirds distill seawater using countercurrent exchange in a gland with a rete mirabile. The gland secretes highly concentrated brine stored near the nostrils above the beak. The bird then "sneezes" the brine out. As freshwater is not usually available in their environments, some seabirds, such as pelicans, petrels, albatrosses, gulls and terns, possess this gland, which allows them to drink the salty water from their environments while they are far from land.

Mangrove trees grow in seawater; they secrete salt by trapping it in parts of the root, which are then eaten by animals (usually crabs). Additional salt is removed by storing it in leaves that fall off. Some types of mangroves have glands on their leaves, which work in a similar way to the seabird desalination gland. Salt is extracted to the leaf exterior as small crystals, which then fall off the leaf.

Willow trees and reeds absorb salt and other contaminants, effectively desalinating the water. This is used in artificial constructed wetlands, for treating sewage.

Sub-field of Desalination

Reverse Osmosis

Water Desalination

Methods

- Distillation
    - Multi-stage flash distillation (MSF)

- o   Multiple-effect distillation (MED | ME)

- o   Vapor-compression (VC)

- Ion exchange

- Membrane processes

    - o   Electrodialysis reversal (EDR)

    - o   Reverse osmosis (RO)

    - o   Nanofiltration (NF)

    - o   Membrane distillation (MD)

    - o   Forward osmosis (FO)

- Freezing desalination

- Geothermal desalination

- Solar desalination

    - o   Solar humidification-Dehumidification (HDH)

    - o   Multiple-effect humidification (MEH)

- Methane hydrate crystallization

- High grade water recycling

Reverse osmosis (RO) is a water purification technology that uses a semipermeable membrane to remove ions, molecules, and larger particles from drinking water. In reverse osmosis, an applied pressure is used to overcome osmotic pressure, a colligative property, that is driven by chemical potential differences of the solvent, a thermodynamic parameter. Reverse osmosis can remove many types of dissolved and suspended species from water, including bacteria, and is used in both industrial processes and the production of potable water. The result is that the solute is retained on the pressurized side of the membrane and the pure solvent is allowed to pass to the other side. To be "selective", this membrane should not allow large molecules or ions through the pores (holes), but should allow smaller components of the solution (such as solvent molecules) to pass freely.

In the normal osmosis process, the solvent naturally moves from an area of low solute concentration (high water potential), through a membrane, to an area of high solute concentration (low water potential). The driving force for the movement of the solvent is the reduction in the free energy of the system when the difference in solvent concentration on either side of a membrane is reduced, generating osmotic pressure due to the solvent moving into the more concentrated solution. Applying an external pressure to reverse the natural flow of pure solvent, thus, is reverse osmosis. The process is similar to other membrane technology applications. However, key differences are found between reverse osmosis and filtration. The predominant removal mechanism in membrane filtration is straining, or size exclusion, so the process can theoretically achieve perfect efficiency regardless of parameters such as the solution's pressure and concentration. Reverse osmosis also involves diffusion, making the process dependent on pressure, flow rate, and other conditions. Reverse osmosis is most commonly known for its use in drinking water purification from seawater, removing the salt and other effluent materials from the water molecules.

History

The process of osmosis through semipermeable membranes was first observed in 1748 by Jean-Antoine Nollet. For the following 200 years, osmosis was only a phenomenon observed in the laboratory. In 1950, the University of California at Los Angeles first investigated desalination of seawater using semipermeable membranes. Researchers from both University of California at Los Angeles and the University of Florida successfully produced fresh water from seawater in the mid-1950s, but the flux was too low to be commercially viable until the discovery at University of California at Los Angeles by Sidney Loeb and Srinivasa Sourirajan at the National Research Council of Canada, Ottawa, of techniques for making asymmetric membranes characterized by an effectively thin "skin" layer supported atop a highly porous and much thicker substrate region of the membrane. John Cadotte, of FilmTec Corporation, discovered that membranes with particularly high flux and low salt passage could be made by interfacial polymerization of m-phenylene diamine and trimesoyl chloride. Cadotte's patent on this process was the subject of litigation and has since expired. Almost all commercial reverse osmosis membrane is now made by this method. By the end of 2001, about 15,200 desalination plants were in operation or in the planning stages, worldwide.

Reverse osmosis production train, North Cape Coral Reverse Osmosis Plant

In 1977 Cape Coral, Florida became the first municipality in the United States to use the RO process on a large scale with an initial operating capacity of 3 million gallons per day. By 1985, due to the rapid growth in population of Cape Coral, the city had the largest low pressure reverse osmosis plant in the world, capable of producing 15 million gallons / day (MGD).

Process

Osmosis is a natural process. When two solutions with different concentrations of a solute are separated by a semipermeable membrane, the solvent has a tendency to move from low to high solute concentrations for chemical potential equilibration.

A semipermeable membrane coil used in desalination

Formally, reverse osmosis is the process of forcing a solvent from a region of high solute concentration through a semipermeable membrane to a region of low solute concentration by applying a pressure in excess of the osmotic pressure. The largest and most important application of reverse osmosis is the separation of pure water from seawater and brackish waters; seawater or brackish water is pressurized against one surface of the membrane, causing transport of salt-depleted water across the membrane and emergence of potable drinking water from the low-pressure side.

The membranes used for reverse osmosis have a dense layer in the polymer matrix—either the skin of an asymmetric membrane or an interfacially polymerized layer within a thin-film-composite membrane—where the separation occurs. In most cases, the membrane is designed to allow only water to pass through this dense layer, while preventing the passage of solutes (such as salt ions). This process requires that a high pressure be exerted on the high concentration side of the membrane, usually 2–17 bar (30–250 psi) for fresh and brackish water, and 40–82 bar (600–1200 psi) for seawater, which has around 27 bar (390 psi) natural osmotic pressure that must be overcome. This process is best known for its use in desalination (removing the salt and other minerals from sea water to get fresh water), but since the early 1970s, it has also been used to purify fresh water for medical, industrial, and domestic applications.

Fresh Water Applications

Drinking Water Purification

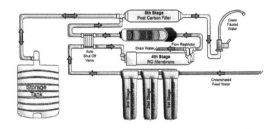

The reverse osmosis water filter process

Around the world, household drinking water purification systems, including a reverse osmosis step, are commonly used for improving water for drinking and cooking.

Such systems typically include a number of steps:

- a sediment filter to trap particles, including rust and calcium carbonate

- optionally, a second sediment filter with smaller pores

- an activated carbon filter to trap organic chemicals and chlorine, which will attack and degrade thin film composite membrane reverse osmosis membranes

- a reverse osmosis filter, which is a thin film composite membrane

- optionally, a second carbon filter to capture those chemicals not removed by the reverse osmosis membrane

- optionally an ultraviolet lamp for sterilizing any microbes that may escape filtering by the reverse osmosis membrane

- latest developments in the sphere include nano materials and membranes

In some systems, the carbon prefilter is omitted, and cellulose triacetate membrane is used. The cellulose triacetate membrane is prone to rotting unless protected by chlorinated water, while the thin film composite membrane is prone to breaking down under the influence of chlorine. In cellulose triacetate membrane systems, a carbon postfilter is needed to remove chlorine from the final product, water.

Portable reverse osmosis water processors are sold for personal water purification in various locations. To work effectively, the water feeding to these units should be under some pressure (40 pounds per square inch (280 kPa) or greater is the norm). Portable reverse osmosis water processors can be used by people who live in rural areas without clean water, far away from the city's water pipes. Rural people filter river or ocean water themselves, as the device is easy to use (saline water may need special membranes). Some travelers on long boating, fishing, or island camping trips, or in countries where the local water supply is polluted or substandard, use reverse osmosis water processors coupled with one or more ultraviolet sterilizers.

In the production of bottled mineral water, the water passes through a reverse osmosis water processor to remove pollutants and microorganisms. In European countries, though, such processing of natural mineral water (as defined by a European Directive) is not allowed under European law. In practice, a fraction of the living bacteria can and do pass through reverse osmosis membranes through minor imperfections, or bypass the membrane entirely through tiny leaks in surrounding seals. Thus, complete reverse osmosis systems may include additional water treatment stages that use ultraviolet light or ozone to prevent microbiological contamination.

Membrane pore sizes can vary from 0.1 to 5,000 nm (4×10^{-9} to 2×10^{-4} in) depending on filter type. Particle filtration removes particles of 1 μm (3.9×10^{-5} in) or larger. Microfiltration removes particles of 50 nm or larger. Ultrafiltration removes particles of roughly 3 nm or larger. Nanofiltration removes particles of 1 nm or larger. Reverse osmosis is in the final category of membrane filtration, hyperfiltration, and removes particles larger than 0.1 nm.

Military Use: The Reverse Osmosis Water Puri ication Unit

United States Marines from Combat Logistics Battalion 31 operate reverse osmosis water purification units for relief efforts after the 2006 Southern Leyte mudslide

A reverse osmosis water purification unit (ROWPU) is a portable, self-contained water treatment plant. Designed for military use, it can provide potable water from nearly any water source. There are many models in use by the United States armed forces and the Canadian Forces. Some models are containerized, some are trailers, and some are vehicles unto themselves.

Each branch of the United States armed forces has their own series of reverse osmosis water purification unit models, but they are all similar. The water is pumped from its raw source into the reverse osmosis water purification unit module, where it is treated with a polymer to initiate coagulation. Next, it is run through a multi-media filter where it undergoes primary treatment by removing turbidity. It is then pumped through a cartridge filter which is usually spiral-wound cotton. This process clarifies the water of any particles larger than 5 micrometres (0.00020 in) and eliminates almost all turbidity.

The clarified water is then fed through a high-pressure piston pump into a series of vessels where it is subject to reverse osmosis. The product water is free of 90.00–99.98% of the raw water's total dissolved solids and by military standards, should have no more than 1000–1500 parts per million by measure of electrical conductivity. It is then disinfected with chlorine and stored for later use.

Within the United States Marine Corps, the reverse osmosis water purification unit has been replaced by both the Lightweight Water Purification System and Tactical Water Purification Systems. The Lightweight Water Purification Systems can be transported by Humvee and filters 125 US gallons (470 l) per hour. The Tactical Water Purification Systems can be carried on a Medium Tactical Vehicle Replacement truck, and can filter 1,200 to 1,500 US gallons (4,500 to 5,700 l) per hour.

Water and Wastewater Purification

Rain water collected from storm drains is purified with reverse osmosis water processors and used for landscape irrigation and industrial cooling in Los Angeles and other cities, as a solution to the problem of water shortages.

In industry, reverse osmosis removes minerals from boiler water at power plants. The water is distilled multiple times. It must be as pure as possible so it does not leave deposits on the machinery or cause corrosion. The deposits inside or outside the boiler tubes may result in underperformance of the boiler, bringing down its efficiency and resulting in poor steam production, hence poor power production at the turbine.

It is also used to clean effluent and brackish groundwater. The effluent in larger volumes (more than 500 m³/d) should be treated in an effluent treatment plant first, and then the clear effluent is subjected to reverse osmosis system. Treatment cost is reduced significantly and membrane life of the reverse osmosis system is increased.

The process of reverse osmosis can be used for the production of deionized water.

Reverse osmosis process for water purification does not require thermal energy. Flow-through reverse osmosis systems can be regulated by high-pressure pumps. The recovery of purified water depends upon various factors, including membrane sizes, membrane pore size, temperature, operating pressure, and membrane surface area.

In 2002, Singapore announced that a process named NEWater would be a significant part of its future water plans. It involves using reverse osmosis to treat domestic wastewater before discharging the NEWater back into the reservoirs.

Food Industry

In addition to desalination, reverse osmosis is a more economical operation for concentrating food liquids (such as fruit juices) than conventional heat-treatment processes. Research has been done on concentration of orange juice and tomato juice. Its advantages include a lower operating cost and the ability to avoid heat-treatment processes, which makes it suitable for heat-sensitive substances such as the protein and enzymes found in most food products.

Reverse osmosis is extensively used in the dairy industry for the production of whey protein powders and for the concentration of milk to reduce shipping costs. In whey applications, the whey (liquid remaining after cheese manufacture) is concentrated with reverse osmosis from 6% total solids to 10–20% total solids before ultrafiltration processing. The ultrafiltration retentate can then be used to make various whey powders, including whey protein isolate. Additionally, the ultrafiltration permeate, which contains lactose, is concentrated by reverse osmosis from 5% total solids to 18–22% total solids to reduce crystallization and drying costs of the lactose powder.

Although use of the process was once avoided in the wine industry, it is now widely understood and used. An estimated 60 reverse osmosis machines were in use in Bordeaux, France, in 2002. Known users include many of the elite classed growths (Kramer) such as Château Léoville-Las Cases in Bordeaux.

Maple Syrup Production

In 1946, some maple syrup producers started using reverse osmosis to remove water from sap before the sap is boiled down to syrup. The use of reverse osmosis allows about 75-90% of the water to be removed from the sap, reducing energy consumption and exposure of the syrup to high temperatures. Microbial contamination and degradation of the membranes must be monitored.

Hydrogen Production

For small-scale hydrogen production, reverse osmosis is sometimes used to prevent formation of minerals on the surface of electrodes.

Reef Aquariums

Many reef aquarium keepers use reverse osmosis systems for their artificial mixture of seawater. Ordinary tap water can contain excessive chlorine, chloramines, copper, nitrates, nitrites, phosphates, silicates, or many other chemicals detrimental to the sensitive organisms in a reef environment. Contaminants such as nitrogen compounds and phosphates can lead to excessive and unwanted algae growth. An effective combination of both reverse osmosis and deionization is the most popular among reef aquarium keepers, and is preferred above other water purification processes due to the low cost of ownership and minimal operating costs. Where chlorine and chloramines are found in the water, carbon filtration is needed before the membrane, as the common residential membrane used by reef keepers does not cope with these compounds.

Window Cleaning

An increasingly popular method of cleaning windows is the so-called "water-fed pole" system. Instead of washing the windows with detergent in the conventional way, they are scrubbed with highly purified water, typically containing less than 10 ppm dissolved solids, using a brush on the end of a long pole which is wielded from ground level. Reverse osmosis is commonly used to purify the water.

Landfill Leachate Purification

Treatment with reverse osmosis is limited, resulting in low recoveries on high concentration (measured with electrical conductivity) and fouling of the RO membranes. Reverse osmosis applicability is limited by conductivity, organics, and scaling inorganic elements such as $CaSO_4$, Si, Fe and Ba. Low organic scaling can be used two different technology, one is using spiral wound membrane type of module, and for high organic scaling, high conductivity and higher pressure (up to 90 bars) can be used disc tube module with reverse osmosis membranes. Disc tube modules was redesigned for landfill leachate purification, what usually is contaminated with high organics. Due to the cross-flow with high velocity is given by a flow booster pump, what is recirculating the flow over the same membrane surface between 1,5 and 3 times before is released as a concentrate. High velocity is also good against membrane scaling and successful membrane cleanings.

Power Consumption for a Disc Tube Module System

disc tube module   spiral wound module

Disc tube module with RO membrane cushion and Spiral wound module with RO membrane

energy consumption per m³ leachate			
name of module	1-stage up to 75 bar	2-stage up to 75 bar	3-stage up to 120 bar
disc tube module	6.1 - 8.1 kWh/m³	8.1 - 9.8 kWh/m³	11.2 - 14.3 kWh/m³

Desalination

Areas that have either no or limited surface water or groundwater may choose to desalinate. Reverse osmosis is an increasingly common method of desalination, because of its relatively low energy consumption. In recent years, energy consumption has dropped to around 3 kWh/m³, with the development of more efficient energy recovery devices and improved membrane materials. According to the International Desalination Association, for 2011, reverse osmosis was used in 66% of installed desalination capacity (44.5 of 67.4 Mm³/day), and nearly all new plants. Other plants mainly use thermal distillation methods: multiple-effect distillation and multi-stage flash.

Sea water reverse osmosis (SWRO) desalination, a membrane process, has been commercially used since the early 1970s. Its first practical use was demonstrated by Sidney Loeb from University of California at Los Angeles in Coalinga, California, and Srinivasa Sourirajan of National Research council, Canada. Because no heating or phase changes are needed, energy requirements are low, around 3 kWh/m³, in comparison to other processes of desalination, but are still much higher than those required for other forms of water supply, including reverse osmosis treatment of wastewater, at 0.1 to 1 kWh/m³. Up to 50% of the seawater input can be recovered as fresh water, though lower recoveries may reduce membrane fouling and energy consumption.

Brackish water reverse osmosis refers to desalination of water with a lower salt content than sea water, usually from river estuaries or saline wells. The process is substantially the same as sea water reverse osmosis, but requires lower pressures and therefore less energy. Up to 80% of the feed water input can be recovered as fresh water, depending on feed salinity.

The Ashkelon sea water reverse osmosis desalination plant in Israel is the largest in the world. The project was developed as a build-operate-transfer by a consortium of three international companies: Veolia water, IDE Technologies, and Elran.

The typical single-pass sea water reverse osmosis system consists of:

- Intake
- Pretreatment
- High pressure pump (if not combined with energy recovery)
- Membrane assembly
- Energy recovery (if used)
- Remineralisation and pH adjustment
- Disinfection
- Alarm/control panel

Pretreatment

Pretreatment is important when working with reverse osmosis and nanofiltration membranes due to the nature of their spiral-wound design. The material is engineered in such a fashion as to allow only one-way flow through the system. As such, the spiral-wound design does not allow for backpulsing with water or air agitation to scour its surface and remove solids. Since accumulated material cannot be removed from the membrane surface systems, they are highly susceptible to fouling (loss of production capacity). Therefore, pretreatment is a necessity for any reverse osmosis or nanofiltration system. Pretreatment in sea water reverse osmosis systems has four major components:

- Screening of solids: Solids within the water must be removed and the water treated to prevent fouling of the membranes by fine particle or biological growth, and reduce the risk of damage to high-pressure pump components.

- Cartridge filtration: Generally, string-wound polypropylene filters are used to remove particles of 1–5 μm diameter.

- Dosing: Oxidizing biocides, such as chlorine, are added to kill bacteria, followed by bisulfite dosing to deactivate the chlorine, which can destroy a thin-film composite membrane. There are also biofouling inhibitors, which do not kill bacteria, but simply prevent them from growing slime on the membrane surface and plant walls.

- Prefiltration pH adjustment: If the pH, hardness and the alkalinity in the feedwater result in a scaling tendency when they are concentrated in the reject stream, acid is dosed to maintain carbonates in their soluble carbonic acid form.

$$CO_3^{2-} + H_3O^+ = HCO_3^- + H_2O$$

$$HCO_3^- + H_3O^+ = H_2CO_3 + H_2O$$

- Carbonic acid cannot combine with calcium to form calcium carbonate scale. Calcium carbonate scaling tendency is estimated using the Langelier saturation index. Adding too much sulfuric acid to control carbonate scales may result in calcium sulfate, barium sulfate, or strontium sulfate scale formation on the reverse osmosis membrane.

- Prefiltration antiscalants: Scale inhibitors (also known as antiscalants) prevent formation of all scales compared to acid, which can only prevent formation of calcium carbonate and calcium phosphate scales. In addition to inhibiting carbonate and phosphate scales, antiscalants inhibit sulfate and fluoride scales and disperse colloids and metal oxides. Despite claims that antiscalants can inhibit silica formation, no concrete evidence proves that silica polymerization can be inhibited by antiscalants. Antiscalants can control acid-soluble scales at a fraction of the dosage required to control the same scale using sulfuric acid.

- Some small scale desalination units use 'beach wells'; they are usually drilled on the seashore in close vicinity to the ocean. These intake facilities are relatively simple to build and the seawater they collect is pretreated via slow filtration through the subsurface sand/seabed formations in the area of source water extraction. Raw seawater collected using beach wells is often of better quality in terms of solids, silt, oil and grease, natural organic contamination and aquatic microorganisms, compared to open seawater intakes. Sometimes, beach intakes may also yield source water of lower salinity.

High Pressure Pump

The high pressure pump supplies the pressure needed to push water through the membrane, even as the membrane rejects the passage of salt through it. Typical pressures for brackish water range from 225 to 376 psi (15.5 to 26 bar, or 1.6 to 2.6 MPa). In the case of seawater, they range from 800 to 1,180 psi (55 to 81.5 bar or 6 to 8 MPa). This requires a large amount of energy. Where energy recovery is used, part of the high pressure pump's work is done by the energy recovery device, reducing the system energy inputs.

Membrane Assembly

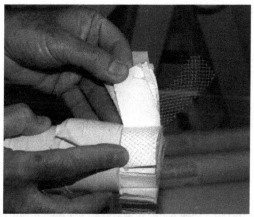
The layers of a membrane

The membrane assembly consists of a pressure vessel with a membrane that allows feedwater to be pressed against it. The membrane must be strong enough to withstand whatever pressure is applied against it. Reverse osmosis membranes are made in a variety of configurations, with the two most common configurations being spiral-wound and hollow-fiber.

Only a part of the saline feed water pumped into the membrane assembly passes through the membrane with the salt removed. The remaining "concentrate" flow passes along the saline side of the membrane to flush away the concentrated salt solution. The percentage of desalinated water produced versus the saline water feed flow is known as the "recovery ratio". This varies with the salinity of the feed water and the system design parameters: typically 20% for small seawater systems, 40% - 50% for larger seawater systems, and 80% - 85% for brackish water. The concentrate flow is at typically only 3 bar / 50 psi less than the feed pressure, and thus still carries much of the high pressure pump input energy.

The desalinated water purity is a function of the feed water salinity, membrane selection and recovery ratio. To achieve higher purity a second pass can be added which generally requires re-pumping. Purity expressed as total dissolved solids typically varies from 100 to 400 parts per million (ppm or milligram/litre)on a seawater feed. A level of 500 ppm is generally accepted as the upper limit for drinking water, while the US Food and Drug Administration classifies mineral water as water containing at least 250 ppm.

Energy Recovery

Energy recovery can reduce energy consumption by 50% or more. Much of the high pressure pump input energy can be recovered from the concentrate flow, and the increasing efficiency of energy recovery devices has greatly reduced the energy needs of reverse osmosis desalination. Devices used, in order of invention, are:

- Turbine or Pelton wheel: a water turbine driven by the concentrate flow, connected to the high pressure pump drive shaft to provide part of its input power. Positive displacement axial piston motors have also been used in place of turbines on smaller systems.

- Turbocharger: a water turbine driven by the concentrate flow, directly connected to a centrifugal pump which boosts the high pressure pump output pressure, reducing the pressure needed from the high pressure pump and thereby its energy input, similar in construction principle to car engine turbochargers.

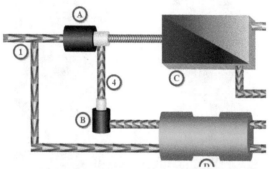

Schematics of a reverse osmosis desalination system using a pressure exchanger.*1*: Sea water inflow,*2*: Fresh water flow (40%),*3*: Concentrate flow (60%),*4*: Sea water flow (60%),*5*: Concentrate (drain),*A: Pump flow (40%),*B: Circulation pump,C: Osmosis unit with membrane,D: Pressure exchanger*

- Pressure exchanger: using the pressurized concentrate flow, in direct contact or via a piston, to pressurize part of the membrane feed flow to near concentrate flow pressure. A boost pump then raises this pressure by typically 3 bar / 50 psi to the membrane feed pressure. This reduces flow needed from the high-pressure pump by an amount equal to the concentrate flow, typically 60%, and thereby its energy input. These are widely used on larger low-energy systems. They are capable of 3 kWh/m³ or less energy consumption.

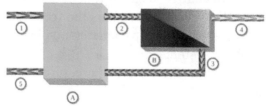

Schematic of a reverse osmosis desalination system using an energy recovery pump.*1*: Sea water inflow (100%, 1 bar),*2*: Sea water flow (100%, 50 bar),*3*: Concentrate flow (60%, 48 bar),*4*: Fresh water flow (40%, 1 bar),*5*: Concentrate to drain (60%,1 bar),*A: Pressure recovery pump,B: Osmosis unit with membrane*

- Energy recovery pump: a reciprocating piston pump having the pressurized concentrate flow applied to one side of each piston to help drive the membrane feed flow from the opposite side. These are the simplest energy recovery devices to apply, combining the high pressure pump and energy recovery in a single self-regulating unit. These are widely used on smaller low-energy systems. They are capable of 3 kWh/m³ or less energy consumption.

Remineralisation and pH Adjustment

The desalinated water is "stabilized" to protect downstream pipelines and storage, usually by adding lime or caustic to prevent corrosion of concrete-lined surfaces. Liming material is used to adjust pH between 6.8 and 8.1 to meet the potable water specifications, primarily for effective disinfection and for corrosion control. Remineralisation may be needed to replace minerals removed from the water by desalination. Although this process has proved to be costly and not very convenient if it is intended to meet mineral demand by humans and plants. The very same mineral demand that freshwater sources provided previously. For instance water from Israel's national water carrier typically contains dissolved magnesium levels of 20 to 25 mg/liter, while water from the Ashkelon plant has no magnesium. After farmers used this water, magnesium deficiency symptoms appeared in crops, including tomatoes, basil, and flowers, and had to be remedied by fertilization. Current Israeli drinking water standards set a minimum calcium level of 20 mg/liter. The postdesalination treatment in the Ashkelon plant uses sulfuric acid to dissolve calcite (limestone), resulting in calcium concentration of 40 to 46 mg/liter. This is still lower than the 45 to 60 mg/liter found in typical Israeli freshwaters.

Disinfection

Post-treatment consists of preparing the water for distribution after filtration. Reverse osmosis is an effective barrier to pathogens, but post-treatment provides secondary protection against compromised membranes and downstream problems. Disinfection by means of ultra violet (UV) lamps (sometimes called germicidal or bactericidal) may be employed to sterilize pathogens which bypassed the reverse osmosis process. Chlorination or chloramination (chlorine and ammonia) protects against pathogens which may have lodged in the distribution system downstream, such as from new construction, backwash, compromised pipes, etc.

Disadvantages

Household reverse osmosis units use a lot of water because they have low back pressure. As a result, they recover only 5 to 15% of the water entering the system. The remainder is discharged as waste water. Because waste water carries with it the rejected contaminants, methods to recover this water are not practical for household systems. Wastewater is typically connected to the house drains and will add to the load on the household septic system. A reverse osmosis unit delivering five gallons of treated water per day may discharge between 20 and 90 gallons of waste water per day.

Large-scale industrial/municipal systems recover typically 75% to 80% of the feed water, or as high as 90%, because they can generate the high pressure needed for higher recovery reverse osmosis filtration. On the other hand, as recovery of wastewater increases in commercial operations, effective contaminant removal rates tend to become reduced, as evidenced by product water total dissolved solids levels.

Due to its fine membrane construction, reverse osmosis not only removes harmful contaminants present in the water, but it also may remove many of the desirable minerals from the water. A number of peer-reviewed studies have looked at the long-term health effects of drinking demineralized water.

Waste Stream Considerations

Depending upon the desired product, either the solvent or solute stream of reverse osmosis will be waste. For food concentration applications, the concentrated solute stream is the product and the solvent stream is waste. For water treatment applications, the solvent stream is purified water and the solute stream is concentrated waste. The solvent waste stream from food processing may be used as reclaimed water, but there may be fewer options for disposal of a concentrated waste solute stream. Ships may use marine dumping and coastal desalination plants typically use marine outfalls. Landlocked reverse osmosis plants may require evaporation ponds or injection wells to avoid polluting groundwater or surface runoff.

New Developments

Since the 1970s, prefiltration of high-fouling waters with another larger-pore membrane, with less hydraulic energy requirement, has been evaluated and sometimes used. However, this means that the water passes through two membranes and is often repressurized, which requires more energy to be put into the system, and thus increases the cost.

Other recent developmental work has focused on integrating reverse osmosis with electrodialysis to improve recovery of valuable deionized products, or to minimize the volume of concentrate requiring discharge or disposal.

In the production of drinking water, the latest developments include nanoscale and graphene membranes.

The world's largest RO desalination plant was built in Sorek, Israel in 2013. It has an output of 624,000 m³/day. It is also the cheapest and will sell water to the authorities for U\$0.58/kL.

Distillation

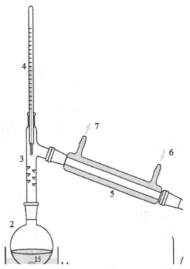

Laboratory display of distillation: 1: A source of heat 2: Still pot 3: Still head 4: Thermometer/Boiling point temperature 5: Condenser 6: Cooling water in 7: Cooling water out 8: Distillate/receiving flask 9: Vacuum/gas inlet 10: Still receiver 11: Heat control 12: Stirrer speed control 13: Stirrer/heat plate 14: Heating (Oil/sand) bath 15: Stirring means e.g. (shown), boiling chips or mechanical stirrer 16: Cooling bath.

Distillation is a process of separating the component substances from a liquid mixture by selective evaporation and condensation. Distillation may result in essentially complete separation (nearly pure components), or it may be a partial separation that increases the concentration of selected components of the mixture. In either case the process exploits differences in the volatility of mixture's components. In industrial chemistry, distillation is a unit operation of practically universal importance, but it is a physical separation process and not a chemical reaction.

Commercially, distillation has many applications. For example:

- In the fossil fuel industry distillation is a major class of operation in obtaining materials from crude oil for fuels and for chemical feedstocks.

- Distillation permits separation of air into its components — notably oxygen, nitrogen, and argon — for industrial use.

- In the field of industrial chemistry, large ranges of crude liquid products of chemical synthesis are distilled to separate them, either from other products, or from impurities, or from unreacted starting materials.

- Distillation of fermented products produces distilled beverages with a high alcohol content, or separates out other fermentation products of commercial value.

An installation for distillation, especially of alcohol, is a distillery. The distillation equipment is a still.

History

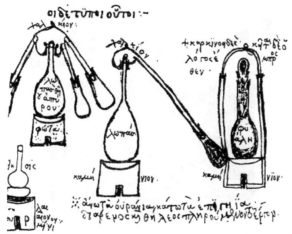

Distillation equipment used by the 3rd century Greek alchemist Zosimos of Panopolis, from the Byzantine Greek manuscript *Parisinus graces*.

Aristotle wrote about the process in his *Meteorologica* and even that "ordinary wine possesses a kind of exhalation, and that is why it gives out a flame". Later evidence of distillation comes from Greek alchemists working in Alexandria in the 1st century AD. Distilled water has been known since at least c. 200, when Alexander of Aphrodisias described the process. Distillation in China could have begun during the Eastern Han Dynasty (1st–2nd centuries), but archaeological evidence indicates that actual distillation of beverages began in the Jin and Southern

Song dynasties. A still was found in an archaeological site in Qinglong, Hebei province dating to the 12th century. Distilled beverages were more common during the Yuan dynasty. Arabs learned the process from the Alexandrians and used it extensively in their chemical experiments.

Clear evidence of the distillation of alcohol comes from the School of Salerno in the 12th century. Fractional distillation was developed by Tadeo Alderotti in the 13th century.

In 1500, German alchemist Hieronymus Braunschweig published *Liber de arte destillandi* (The Book of the Art of Distillation) the first book solely dedicated to the subject of distillation, followed in 1512 by a much expanded version. In 1651, John French published The Art of Distillation the first major English compendium of practice, though it has been claimed that much of it derives from Braunschweig's work. This includes diagrams with people in them showing the industrial rather than bench scale of the operation.

Hieronymus Brunschwig's *Liber de arte Distillandi de Compositis* (Strassburg, 1512) Chemical Heritage Foundation

A retort

Distillation

Old Ukrainian vodka still

Simple liqueur distillation in East Timor

As alchemy evolved into the science of chemistry, vessels called retorts became used for distillations. Both alembics and retorts are forms of glassware with long necks pointing to the side at a downward angle which acted as air-cooled condensers to condense the distillate and let it drip downward for collection. Later, copper alembics were invented. Riveted joints were often kept tight by using various mixtures, for instance a dough made of rye flour. These alembics often featured a cooling system around the beak, using cold water for instance, which made the condensation of alcohol more efficient. These were called pot stills. Today, the retorts and pot stills have been largely supplanted by more efficient distillation methods in most industrial processes. However, the pot still is still widely used for the elaboration of some fine alcohols such as cognac, Scotch whisky, tequila and some vodkas. Pot stills made of various materials (wood, clay, stainless steel) are also used by bootleggers in various countries. Small pot stills are also sold for the domestic production of flower water or essential oils.

Early forms of distillation were batch processes using one vaporization and one condensation. Purity was improved by further distillation of the condensate. Greater volumes were processed by simply repeating the distillation. Chemists were reported to carry out as many as 500 to 600 distillations in order to obtain a pure compound.

In the early 19th century the basics of modern techniques including pre-heating and reflux were developed, particularly by the French, then in 1830 a British Patent was issued to Aeneas Coffey for a whisky distillation column, which worked continuously and may be regarded as the archetype of modern petrochemical units. In 1877, Ernest Solvay was granted a U.S. Patent for a tray column for ammonia distillation and the same and subsequent years saw developments of this theme for oil and spirits.

With the emergence of chemical engineering as a discipline at the end of the 19th century, scientific rather than empirical methods could be applied. The developing petroleum industry in the early 20th century provided the impetus for the development of accurate design methods such as the McCabe–Thiele method and the Fenske equation. The availability of powerful computers has also allowed direct computer simulation of distillation columns.

Applications of Distillation

The application of distillation can roughly be divided in four groups: laboratory scale, industrial distillation, distillation of herbs for perfumery and medicinals (herbal distillate), and food processing. The latter two are distinctively different from the former two in that in the processing of beverages and herbs, the distillation is not used as a true purification method but more to transfer all volatiles from the source materials to the distillate.

The main difference between laboratory scale distillation and industrial distillation is that laboratory scale distillation is often performed batch-wise, whereas industrial distillation often occurs continuously. In batch distillation, the composition of the source material, the vapors of the distilling compounds and the distillate change during the distillation. In batch distillation, a still is charged (supplied) with a batch of feed mixture, which is then separated into its component fractions which are collected sequentially from most volatile to less volatile, with the bottoms (remaining least or non-volatile fraction) removed at the end. The still can then be recharged and the process repeated.

In continuous distillation, the source materials, vapors, and distillate are kept at a constant composition by carefully replenishing the source material and removing fractions from both vapor and liquid in the system. This results in a better control of the separation process.

Idealized Distillation Model

The boiling point of a liquid is the temperature at which the vapor pressure of the liquid equals the pressure around the liquid, enabling bubbles to form without being crushed. A special case is the normal boiling point, where the vapor pressure of the liquid equals the ambient atmospheric pressure.

It is a common misconception that in a liquid mixture at a given pressure, each component boils at the boiling point corresponding to the given pressure and the vapors of each component will collect separately and purely. This, however, does not occur even in an idealized system. Idealized models of distillation are essentially governed by Raoult's law and Dalton's law, and assume that vapor–liquid equilibria are attained.

Raoult's law states that the vapor pressure of a solution is dependent on 1) the vapor pressure of each chemical component in the solution and 2) the fraction of solution each component makes up a.k.a. the mole fraction. This law applies to ideal solutions, or solutions that have different components but whose molecular interactions are the same as or very similar to pure solutions.

Dalton's law states that the total pressure is the sum of the partial pressures of each individual component in the mixture. When a multi-component liquid is heated, the vapor pressure of each component will rise, thus causing the total vapor pressure to rise. When the total vapor pressure reaches the pressure surrounding the liquid, boiling occurs and liquid turns to gas throughout the bulk of the liquid. Note that a mixture with a given composition has one boiling point at a given pressure, when the components are mutually soluble. A mixture of constant composition does not have multiple boiling points.

An implication of one boiling point is that lighter components never cleanly "boil first". At boiling point, all volatile components boil, but for a component, its percentage in the vapor is the same as its percentage of the total vapor pressure. Lighter components have a higher partial pressure and thus are concentrated in the vapor, but heavier volatile components also have a (smaller) partial pressure and necessarily evaporate also, albeit being less concentrated in the vapor. Indeed, batch distillation and fractionation succeed by varying the composition of the mixture. In batch distillation, the batch evaporates, which changes its composition; in fractionation, liquid higher in the fractionation column contains more lights and boils at lower temperatures. Therefore, starting from a given mixture, it appears to have a boiling range instead of a boiling *point*, although this is because its composition changes: each intermediate mixture has its own, singular boiling point.

The idealized model is accurate in the case of chemically similar liquids, such as benzene and toluene. In other cases, severe deviations from Raoult's law and Dalton's law are observed, most famously in the mixture of ethanol and water. These compounds, when heated together, form an azeotrope, which is a composition with a boiling point higher or lower than the boiling point of each separate liquid. Virtually all liquids, when mixed and heated, will display azeotropic behaviour. Although there are computational methods that can be used to estimate the behavior of a mixture of arbitrary components, the only way to obtain accurate vapor–liquid equilibrium data is by measurement.

It is not possible to *completely* purify a mixture of components by distillation, as this would require each component in the mixture to have a zero partial pressure. If ultra-pure products are the goal, then further chemical separation must be applied. When a binary mixture is evaporated and the other component, e.g. a salt, has zero partial pressure for practical purposes, the process is simpler and is called evaporation in engineering.

Batch Distillation

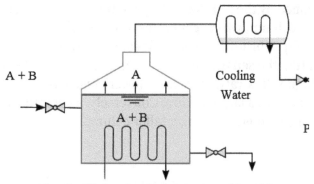

A batch still showing the separation of A and B.

Heating an ideal mixture of two volatile substances A and B (with A having the higher volatility, or lower boiling point) in a batch distillation setup (such as in an apparatus depicted in the opening figure) until the mixture is boiling results in a vapor above the liquid which contains a mixture of A and B. The ratio between A and B in the vapor will be different from the ratio in the liquid: the ratio in the liquid will be determined by how the original mixture was prepared, while the ratio in the vapor will be enriched in the more volatile compound, A (due to Raoult's Law). The vapor goes through the condenser and is removed from the system. This in turn means that the ratio of compounds in the remaining liquid is now different from the initial ratio (i.e., more enriched in B than the starting liquid).

The result is that the ratio in the liquid mixture is changing, becoming richer in component B. This causes the boiling point of the mixture to rise, which in turn results in a rise in the temperature in the vapor, which results in a changing ratio of A : B in the gas phase (as distillation continues, there is an increasing proportion of B in the gas phase). This results in a slowly changing ratio A : B in the distillate.

If the difference in vapor pressure between the two components A and B is large (generally expressed as the difference in boiling points), the mixture in the beginning of the distillation is highly enriched in component A, and when component A has distilled off, the boiling liquid is enriched in component B.

Continuous Distillation

Continuous distillation is an ongoing distillation in which a liquid mixture is continuously (without interruption) fed into the process and separated fractions are removed continuously as output streams occur over time during the operation. Continuous distillation produces a minimum of two output fractions, including at least one volatile distillate fraction, which has boiled and been separately captured as a vapor, and then condensed to a liquid. There is always a bottoms (or residue) fraction, which is the least volatile residue that has not been separately captured as a condensed vapor.

Continuous distillation differs from batch distillation in the respect that concentrations should not change over time. Continuous distillation can be run at a steady state for an arbitrary amount of time. For any source material of specific composition, the main variables that affect the purity of products in continuous distillation are the reflux ratio and the number of theoretical equilibrium stages, in practice determined by the number of trays or the height of packing. Reflux is a flow from the condenser back to the column, which generates a recycle that allows a better separation with a given number of trays. Equilibrium stages are ideal steps where compositions achieve vapor–liquid equilibrium, repeating the separation process and allowing better separation given a reflux ratio. A column with a high reflux ratio may have fewer stages, but it refluxes a large amount of liquid, giving a wide column with a large holdup. Conversely, a column with a low reflux ratio must have a large number of stages, thus requiring a taller column.

General Improvements

Both batch and continuous distillations can be improved by making use of a fractionating column on top of the distillation flask. The column improves separation by providing a larger surface area for the vapor and condensate to come into contact. This helps it remain at equilibrium for as long as possible. The column can even consist of small subsystems ('trays' or 'dishes') which all contain an enriched, boiling liquid mixture, all with their own vapor–liquid equilibrium.

There are differences between laboratory-scale and industrial-scale fractionating columns, but the principles are the same. Examples of laboratory-scale fractionating columns (in increasing efficiency) include

- Air condenser

- Vigreux column (usually laboratory scale only)

- Packed column (packed with glass beads, metal pieces, or other chemically inert material)

- Spinning band distillation system.

Laboratory Scale Distillation

Laboratory scale distillations are almost exclusively run as batch distillations. The device used in distillation, sometimes referred to as a *still*, consists at a minimum of a reboiler or *pot* in which the source material is heated, a condenser in which the heated vapour is cooled back to the liquid state, and a receiver in which the concentrated or purified liquid, called the distillate, is collected. Several laboratory scale techniques for distillation exist.

Typical laboratory distillation unit

Simple Distillation

In simple distillation, the vapor is immediately channeled into a condenser. Consequently, the distillate is not pure but rather its composition is identical to the composition of the vapors at the given temperature and pressure. That concentration follows Raoult's law.

As a result, simple distillation is effective only when the liquid boiling points differ greatly (rule of thumb is 25 °C) or when separating liquids from non-volatile solids or oils. For these cases, the vapor pressures of the components are usually different enough that the distillate may be sufficiently pure for its intended purpose.

Fractional Distillation

For many cases, the boiling points of the components in the mixture will be sufficiently close that Raoult's law must be taken into consideration. Therefore, fractional distillation must be used in order to separate the components by repeated vaporization-condensation cycles within a packed fractionating column. This separation, by successive distillations, is also referred to as rectification.

As the solution to be purified is heated, its vapors rise to the fractionating column. As it rises, it cools, condensing on the condenser walls and the surfaces of the packing material. Here, the condensate continues to be heated by the rising hot vapors; it vaporizes once more. However, the composition of the fresh vapors are determined once again by Raoult's law. Each vaporization-condensation cycle (called a *theoretical plate*) will yield a purer solution of the more volatile component. In reality, each cycle at a given temperature does not occur at exactly the same position in the fractionating column; *theoretical plate* is thus a concept rather than an accurate description.

More theoretical plates lead to better separations. A spinning band distillation system uses a spinning band of Teflon or metal to force the rising vapors into close contact with the descending condensate, increasing the number of theoretical plates.

Steam Distillation

Like vacuum distillation, steam distillation is a method for distilling compounds which are heat-sensitive. The temperature of the steam is easier to control than the surface of a heating element, and allows a high rate of heat transfer without heating at a very high temperature. This process involves bubbling steam through a heated mixture of the raw material. By Raoult's law, some of the target compound will vaporize (in accordance with its partial pressure). The vapor mixture is cooled and condensed, usually yielding a layer of oil and a layer of water.

Steam distillation of various aromatic herbs and flowers can result in two products; an essential oil as well as a watery herbal distillate. The essential oils are often used in perfumery and aromatherapy while the watery distillates have many applications in aromatherapy, food processing and skin care.

Dimethyl sulfoxide usually boils at 189 °C. Under a vacuum, it distills off into the receiver at only 70 °C.

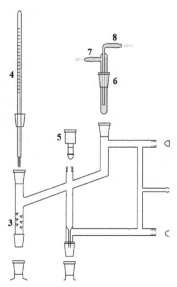

Perkin triangle distillation setup1: Stirrer bar/anti-bumping granules 2: Still pot 3: Fractionating column 4: Thermometer/Boiling point temperature 5: Teflon tap 1 6: Cold finger 7: Cooling water out 8: Cooling water in 9: Teflon tap 2 10: Vacuum/gas inlet 11: Teflon tap 3 12: Still receiver

Vacuum Distillation

Some compounds have very high boiling points. To boil such compounds, it is often better to lower the pressure at which such compounds are boiled instead of increasing the temperature. Once the pressure is lowered to the vapor pressure of the compound (at the given temperature), boiling and the rest of the distillation process can commence. This technique is referred to as vacuum distillation and it is commonly found in the laboratory in the form of the rotary evaporator.

This technique is also very useful for compounds which boil beyond their decomposition temperature at atmospheric pressure and which would therefore be decomposed by any attempt to boil them under atmospheric pressure.

Molecular distillation is vacuum distillation below the pressure of 0.01 torr. 0.01 torr is one order of magnitude above high vacuum, where fluids are in the free molecular flow regime, i.e. the mean free path of molecules is comparable to the size of the equipment. The gaseous phase no longer exerts significant pressure on the substance to be evaporated, and consequently, rate of evaporation no longer depends on pressure. That is, because the continuum assumptions of fluid dynamics no longer apply, mass transport is governed by molecular dynamics rather than fluid dynamics. Thus, a short path between the hot surface and the cold surface is necessary, typically by suspending a hot plate covered with a film of feed next to a cold plate with a line of sight in between. Molecular distillation is used industrially for purification of oils.

Air-sensitive Vacuum Distillation

Some compounds have high boiling points as well as being air sensitive. A simple vacuum distillation system as exemplified above can be used, whereby the vacuum is replaced with an inert gas after the distillation is complete. However, this is a less satisfactory system if one desires to collect fractions under a reduced pressure. To do this a "cow" or "pig" adaptor can be added to the end of the condenser, or for better results or for very air sensitive compounds a Perkin triangle apparatus can be used.

The Perkin triangle, has means via a series of glass or Teflon taps to allows fractions to be isolated from the rest of the still, without the main body of the distillation being removed from either the vacuum or heat source, and thus can remain in a state of reflux. To do this, the sample is first isolated from the vacuum by means of the taps, the vacuum over the sample is then replaced with an inert gas (such as nitrogen or argon) and can then be stoppered and removed. A fresh collection vessel can then be added to the system, evacuated and linked back into the distillation system via the taps to collect a second fraction, and so on, until all fractions have been collected.

Short Path Distillation

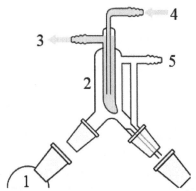

Short path vacuum distillation apparatus with vertical condenser (cold finger), to minimize the distillation path; 1: Still pot with stirrer bar/anti-bumping granules 2: Cold finger – bent to direct condensate 3: Cooling water out 4: cooling water in 5: Vacuum/gas inlet 6: Distillate flask/distillate.

Short path distillation is a distillation technique that involves the distillate travelling a short distance, often only a few centimeters, and is normally done at reduced pressure. A classic example would be a distillation involving the distillate travelling from one glass bulb to another, without the need for a condenser separating the two chambers. This technique is often used for compounds which are unstable at high temperatures or to purify small amounts of compound. The advantage is that the heating temperature can be considerably lower (at reduced pressure) than the boiling point of the liquid at standard pressure, and the distillate only has to travel a short distance before condensing. A short path ensures that little compound is lost on the sides of the apparatus. The Kugelrohr is a kind of a short path distillation apparatus which often contain multiple chambers to collect distillate fractions.

Zone Distillation

Zone distillation is a distillation process in long container with partial melting of refined matter in moving liquid zone and condensation of vapor in the solid phase at condensate pulling in cold area. The process is worked in theory. When zone heater is moving from the top to the bottom of the container then solid condensate with irregular impurity distribution is forming. Then most pure part of the condensate may be extracted as product. The process may be iterated many times by moving (without turnover) the received condensate to the bottom part of the container on the place of refined matter. The irregular impurity distribution in the condensate (that is efficiency of purification) increases with number of repetitions of the process. Zone distillation is a distillation analog of zone recrystallization. Impurity distribution in the condensate is described by known equations of zone recrystallization with various numbers of iteration of process – with replacement distribution efficient k of crystallization on separation factor α of distillation.

Other Types

- The process of reactive distillation involves using the reaction vessel as the still. In this process, the product is usually significantly lower-boiling than its reactants. As the product is formed from the reactants, it is vaporized and removed from the reaction mixture. This technique is an example of a continuous vs. a batch process; advantages include less downtime to charge the reaction vessel with starting material, and less workup. Distillation "over a reactant" could be classified as a reactive distillation. It is typically used to remove volatile impurity from the distallation feed. For example, a little lime may be added to remove carbon dioxide from water followed by a second distillation with a little sulphuric acid added to remove traces of ammonia.

- Catalytic distillation is the process by which the reactants are catalyzed while being distilled to continuously separate the products from the reactants. This method is used to assist equilibrium reactions reach completion.

- Pervaporation is a method for the separation of mixtures of liquids by partial vaporization through a non-porous membrane.

- Extractive distillation is defined as distillation in the presence of a miscible, high boiling, relatively non-volatile component, the solvent, that forms no azeotrope with the other components in the mixture.

- Flash evaporation (or partial evaporation) is the partial vaporization that occurs when a saturated liquid stream undergoes a reduction in pressure by passing through a throttling valve or other throttling device. This process is one of the simplest unit operations, being equivalent to a distillation with only one equilibrium stage.

- Codistillation is distillation which is performed on mixtures in which the two compounds are not miscible.

The unit process of evaporation may also be called "distillation":

- In rotary evaporation a vacuum distillation apparatus is used to remove bulk solvents from a sample. Typically the vacuum is generated by a water aspirator or a membrane pump.

- In a kugelrohr a short path distillation apparatus is typically used (generally in combination with a (high) vacuum) to distill high boiling (> 300 °C) compounds. The apparatus consists of an oven in which the compound to be distilled is placed, a receiving portion which is outside of the oven, and a means of rotating the sample. The vacuum is normally generated by using a high vacuum pump.

Other uses:

- Dry distillation or destructive distillation, despite the name, is not truly distillation, but rather a chemical reaction known as pyrolysis in which solid substances are heated in an inert or reducing atmosphere and any volatile fractions, containing high-boiling liquids and products of pyrolysis, are collected. The destructive distillation of wood to give methanol is the root of its common name – *wood alcohol.*

- Freeze distillation is an analogous method of purification using freezing instead of evaporation. It is not truly distillation, but a recrystallization where the product is the mother liquor, and does not produce products equivalent to distillation. This process is used in the production of ice beer and ice wine to increase ethanol and sugar content, respectively. It is also used to produce applejack. Unlike distillation, freeze distillation concentrates poisonous congeners rather than removing them; As a result, many countries prohibit such applejack as a health measure. However, reducing methanol with the absorption of 4A molecular sieve is a practical method for production. Also, distillation by evaporation can separate these since they have different boiling points.

Azeotropic Distillation

Interactions between the components of the solution create properties unique to the solution, as most processes entail nonideal mixtures, where Raoult's law does not hold. Such interactions can result in a constant-boiling azeotrope which behaves as if it were a pure compound (i.e., boils at a single temperature instead of a range). At an azeotrope, the solution contains the given component in the same proportion as the vapor, so that evaporation does not change the purity, and distillation does not effect separation. For example, ethyl alcohol and water form an azeotrope of 95.6% at 78.1 °C.

If the azeotrope is not considered sufficiently pure for use, there exist some techniques to break the azeotrope to give a pure distillate. This set of techniques are known as azeotropic distillation. Some techniques achieve this by "jumping" over the azeotropic composition (by adding another component to create a new azeotrope, or by varying the pressure). Others work by chemically or physically removing or sequestering the impurity. For example, to purify ethanol beyond 95%, a drying agent (or desiccant, such as potassium carbonate) can be added to convert the soluble water into insoluble water of crystallization. Molecular sieves are often used for this purpose as well.

Immiscible liquids, such as water and toluene, easily form azeotropes. Commonly, these azeotropes are referred to as a low boiling azeotrope because the boiling point of the azeotrope is lower than the boiling point of either pure component. The temperature and composition of the azeotrope is easily predicted from the vapor pressure of the pure components, without use of Raoult's law. The azeotrope is easily broken in a distillation set-up by using a liquid–liquid separator (a decanter) to separate the two liquid layers that are condensed overhead. Only one of the two liquid layers is refluxed to the distillation set-up.

High boiling azeotropes, such as a 20 weight percent mixture of hydrochloric acid in water, also exist. As implied by the name, the boiling point of the azeotrope is greater than the boiling point of either pure component.

To break azeotropic distillations and cross distillation boundaries, such as in the DeRosier Problem, it is necessary to increase the composition of the light key in the distillate.

Breaking an Azeotrope with Unidirectional Pressure Manipulation

The boiling points of components in an azeotrope overlap to form a band. By exposing an azeotrope to a vacuum or positive pressure, it's possible to bias the boiling point of one component away from the other by exploiting the differing vapour pressure curves of each; the curves may overlap at the azeotropic point, but are unlikely to be remain identical further along the pressure axis either side of the azeotropic point. When the bias is great enough, the two boiling points no longer overlap and so the azeotropic band disappears.

This method can remove the need to add other chemicals to a distillation, but it has two potential drawbacks.

Under negative pressure, power for a vacuum source is needed and the reduced boiling points of the distillates requires that the condenser be run cooler to prevent distillate vapours being lost to the vacuum source. Increased cooling demands will often require additional energy and possibly new equipment or a change of coolant.

Alternatively, if positive pressures are required, standard glassware can not be used, energy must be used for pressurization and there is a higher chance of side reactions occurring in the distillation, such as decomposition, due to the higher temperatures required to effect boiling.

A unidirectional distillation will rely on a pressure change in one direction, either positive or negative.

Pressure-swing Distillation

Pressure-swing distillation is essentially the same as the unidirectional distillation used to break azeotropic mixtures, but here both positive and negative pressures may be employed.

This improves the selectivity of the distillation and allows a chemist to optimize distillation by avoiding extremes of pressure and temperature that waste energy. This is particularly important in commercial applications.

One example of the application of pressure-swing distillation is during the industrial purification of ethyl acetate after its catalytic synthesis from ethanol.

Industrial Distillation

Large scale industrial distillation applications include both batch and continuous fractional, vacuum, azeotropic, extractive, and steam distillation. The most widely used industrial applications of continuous, steady-state fractional distillation are in petroleum refineries, petrochemical and chemical plants and natural gas processing plants.

To control and optimize such industrial distillation, a standardized laboratory method, ASTM D86, is established. This test method extends to the atmospheric distillation of petroleum products using a laboratory batch distillation unit to quantitatively determine the boiling range characteristics of petroleum products.

Typical industrial distillation towers

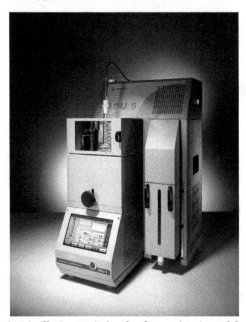

Automatic Distillation Unit for the determination of the boiling
range of petroleum products at atmospheric pressure

Industrial distillation is typically performed in large, vertical cylindrical columns known as distillation towers or distillation columns with diameters ranging from about 65 centimeters to 16 meters and heights ranging from about 6 meters to 90 meters or more. When the process feed has a diverse composition, as in distilling crude oil, liquid outlets at intervals up the column allow for the withdrawal of different *fractions* or products having different boiling points or boiling ranges. The "lightest" products (those with the lowest boiling point) exit from the top of the columns and the "heaviest" products (those with the highest boiling point) exit from the bottom of the column and are often called the bottoms.

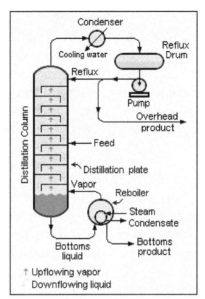

Diagram of a typical industrial distillation tower

Industrial towers use reflux to achieve a more complete separation of products. Reflux refers to the portion of the condensed overhead liquid product from a distillation or fractionation tower that is returned to the upper part of the tower as shown in the schematic diagram of a typical, large-scale industrial distillation tower. Inside the tower, the downflowing reflux liquid provides cooling and condensation of the upflowing vapors thereby increasing the efficiency of the distillation tower. The more reflux that is provided for a given number of theoretical plates, the better the tower's separation of lower boiling materials from higher boiling materials. Alternatively, the more reflux that is provided for a given desired separation, the fewer the number of theoretical plates required. Chemical engineers must choose what combination of reflux rate and number of plates is both economically and physically feasible for the products purified in the distillation column.

Such industrial fractionating towers are also used in cryogenic air separation, producing liquid oxygen, liquid nitrogen, and high purity argon. Distillation of chlorosilanes also enables the production of high-purity silicon for use as a semiconductor.

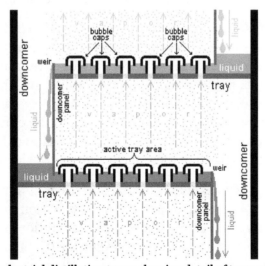

Section of an industrial distillation tower showing detail of trays with bubble caps

Design and operation of a distillation tower depends on the feed and desired products. Given a simple, binary component feed, analytical methods such as the McCabe–Thiele method or the Fenske equation can be used. For a multi-component feed, simulation models are used both for design and operation. Moreover, the efficiencies of the vapor–liquid contact devices (referred to as "plates" or "trays") used in distillation towers are typically lower than that of a theoretical 100% efficient equilibrium stage. Hence, a distillation tower needs more trays than the number of theoretical vapor–liquid equilibrium stages. A variety of models have been postulated to estimate tray efficiencies.

In modern industrial uses, a packing material is used in the column instead of trays when low pressure drops across the column are required. Other factors that favor packing are: vacuum systems, smaller diameter columns, corrosive systems, systems prone to foaming, systems requiring low liquid holdup, and batch distillation. Conversely, factors that favor plate columns are: presence of solids in feed, high liquid rates, large column diameters, complex columns, columns with wide feed composition variation, columns with a chemical reaction, absorption columns, columns limited by foundation weight tolerance, low liquid rate, large turn-down ratio and those processes subject to process surges.

Large-scale, industrial vacuum distillation column

This packing material can either be random dumped packing (1–3" wide) such as Raschig rings or structured sheet metal. Liquids tend to wet the surface of the packing and the vapors pass across this wetted surface, where mass transfer takes place. Unlike conventional tray distillation in which every tray represents a separate point of vapor–liquid equilibrium, the vapor–liquid equilibrium curve in a packed column is continuous. However, when modeling packed columns, it is useful to compute a number of "theoretical stages" to denote the separation efficiency of the packed column with respect to more traditional trays. Differently shaped packings have different surface areas and void space between packings. Both of these factors affect packing performance.

Another factor in addition to the packing shape and surface area that affects the performance of random or structured packing is the liquid and vapor distribution entering the packed bed. The number of theoretical stages required to make a given separation is calculated using a specific vapor to liquid ratio. If the liquid and vapor are not evenly distributed across the superficial tower area as it enters the packed bed, the liquid to vapor ratio will not be correct in the packed bed and the required separation will not be achieved. The packing will appear to not be working properly. The height equivalent to a theoretical plate (HETP) will be greater than expected. The problem is not the packing itself but the mal-distribution of the fluids entering the packed bed. Liquid mal-distribution is more frequently the problem than vapor. The design of the liquid distributors used to introduce the feed and reflux to a packed bed is critical to making the packing perform to it maximum efficiency. Methods of evaluating the effectiveness of a liquid distributor to evenly distribute the liquid entering a packed bed can be found in references. Considerable work as been done on this topic by Fractionation Research, Inc. (commonly known as FRI).

Multi-effect Distillation

The goal of multi-effect distillation is to increase the energy efficiency of the process, for use in desalination, or in some cases one stage in the production of ultrapure water. The number of effects is inversely proportional to the kW·h/m³ of water recovered figure, and refers to the volume of water recovered per unit of energy compared with single-effect distillation. One effect is roughly 636 kW·h/m³.

- Multi-stage flash distillation Can achieve more than 20 effects with thermal energy input, as mentioned in the article.

- Vapor compression evaporation Commercial large-scale units can achieve around 72 effects with electrical energy input, according to manufacturers.

There are many other types of multi-effect distillation processes, including one referred to as simply multi-effect distillation (MED), in which multiple chambers, with intervening heat exchangers, are employed.

Distillation in Food Processing

Distilled Beverages

Carbohydrate-containing plant materials are allowed to ferment, producing a dilute solution of ethanol in the process. Spirits such as whiskey and rum are prepared by distilling these dilute solutions of ethanol. Components other than ethanol, including water, esters, and other alcohols, are collected in the condensate, which account for the flavor of the beverage. Some of these beverages are then stored in barrels or other containers to acquire more flavor compounds and characteristic flavors.

Gallery

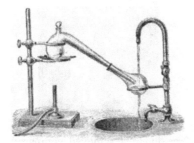

Chemistry in its beginnings used retorts as
laboratory equipment exclusively for distillation processes.

A simple set-up to distill dry and oxygen-free toluene.

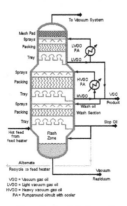

Diagram of an industrial-scale vacuum
distillation column as commonly used in oil refineries

A rotary evaporator is able to distill solvents
more quickly at lower temperatures through the use of a vacuum.

Distillation using semi-microscale apparatus. The jointless design eliminates the need to fit pieces together. The pear-shaped flask allows the last drop of residue to be removed, compared with a similarly-sized round-bottom flask The small holdup volume prevents losses. A pig is used to channel the various distillates into three receiving flasks. If necessary the distillation can be carried out under vacuum using the vacuum adapter at the pig.

References

- Peter Gleick; Heather Cooley; David Katz (2006). The world's water, 2006–2007: the biennial report on freshwater resources. Island Press. pp. 29–31. ISBN 1-59726-106-8. Retrieved 12 September 2009.

- Cooley, Heather; Gleick, Peter H. and Wolff, Gary (June 2006) DESALINATION, WITH A GRAIN OF SALT. A California Perspective, Pacific Institute for Studies in Development, Environment, and Security. ISBN 1-893790-13-4

- Crittenden, John; Trussell, Rhodes; Hand, David; Howe, Kerry and Tchobanoglous, George. Water Treatment Principles and Design, Edition 2. John Wiley and Sons. New Jersey. 2005 ISBN 0-471-11018-3

- Bryan H. Bunch; Alexander Hellemans (2004). The History of Science and Technology. Houghton Mifflin Harcourt. p. 88. ISBN 0-618-22123-9.

- Forbes, Robert James (1970). A short history of the art of distillation: from the beginnings up to the death of Cellier Blumenthal. BRILL. pp. 57, 89. ISBN 978-90-04-00617-1. Retrieved 29 June 2010.

- D. F. Othmer (1982) "Distillation – Some Steps in its Development", in W. F. Furter (ed) A Century of Chemical Engineering ISBN 0-306-40895-3

- Perry, Robert H.; Green, Don W. (1984). Perry's Chemical Engineers' Handbook (6th ed.). McGraw-Hill. ISBN 0-07-049479-7.

- Thiel, Gregory P. (2015-06-01). "Salty solutions". Physics Today. 68 (6): 66–67. Bibcode:2015PhT....68f..66T. doi:10.1063/PT.3.2828. ISSN 0031-9228.

- "Desalination plant powered by waste heat opens in Maldives" European Innovation Partnerships (EIP) news. Retrieved March 18, 2014.

- Cao, Liwei (June 5, 2013). "Dais Analytic Corporation Announces Product Sale to Asia, Functional Waste Water Treatment Pilot, and Key Infrastructure Appointments". PR Newswire. Retrieved July 9, 2013.

- "Sandia National Labs: Desalination and Water Purification: Research and Development". sandia.gov. 2007. Retrieved July 9, 2013.

- "Chemists Work to Desalinate the Ocean for Drinking Water, One Nanoliter at a Time". Science Daily. June 27, 2013. Retrieved June 29, 2013.

4

Understanding Storm Water

To understand storm water elaborately, the following text gives an insight on storm water harvesting. Storm water harvesting is the collection, accumulation, treatment or purification, and storing of storm water for it's eventual reuse. The chapter serves as a source to understand the major categories related to storm water.

Stormwater

Stormwater is water that originates during precipitation events and snow/ice melt. Stormwater can soak into the soil (infiltrate), be held on the surface and evaporate, or runoff and end up in nearby streams, rivers, or other water bodies (surface water).

Urban runoff entering a storm drain

In natural landscapes such as forests, the soil absorbs much of the stormwater and plants help hold stormwater close to where it falls. In developed environments, unmanaged stormwater can create two major issues: one related to the volume and timing of runoff water (flooding) and the other related to potential contaminants that the water is carrying (water pollution).

Stormwater is also a resource and important as the world's human population demand exceeds the availability of readily available water. Techniques of storm water harvesting with point source water management and purification can potentially make urban environments self-sustaining in terms of water.

Stormwater Pollution

Because impervious surfaces (parking lots, roads, buildings, compacted soil) do not allow rain to infiltrate into the ground, more runoff is generated than in the undeveloped condition. This additional runoff can erode watercourses (streams and rivers) as well as cause flooding after the stormwater collection system is overwhelmed by the additional flow. Because the water is flushed out of the watershed during the storm event, little infiltrates the soil, replenishes groundwater, or supplies stream baseflow in dry weather.

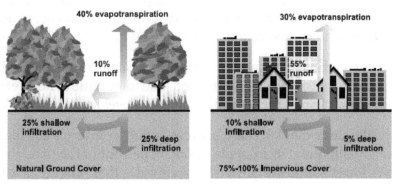

Relationship between impervious surfaces and surface runoff

A first flush is the initial runoff of a rainstorm. During this phase, polluted water entering storm drains in areas with high proportions of impervious surfaces is typically more concentrated compared to the remainder of the storm. Consequently, these high concentrations of urban runoff result in high levels of pollutants discharged from storm sewers to surface waters.

Pollutants entering surface waters during precipitation events is termed polluted runoff. Daily human activities result in deposition of pollutants on roads, lawns, roofs, farm fields, etc. When it rains or there is irrigation, water runs off and ultimately makes its way to a river, lake, or the ocean. While there is some attenuation of these pollutants before entering the receiving waters, the quantity of human activity results in large enough quantities of pollutants to impair these receiving waters.

Stormwater Runoff as a Source of Pollution

Urban runoff being discharged to coastal waters

In addition to the pollutants carried in stormwater runoff, urban runoff is being recognized as a cause of pollution in its own right. In natural catchments (watersheds) surface runoff entering waterways is a relatively rare event, occurring only a few times each year and generally after larger storm events. Before development occurred most rainfall soaked into the ground and contributed to groundwater recharge or was recycled into the atmosphere by vegetation through evapotranspiration .

Modern drainage systems which collect runoff from impervious surfaces (e.g., roofs and roads) ensure that water is efficiently conveyed to waterways through pipe networks, meaning that even small storm events result in increased waterway flows.

In addition to delivering higher pollutants from the urban catchment, increased stormwater flow can lead to stream erosion, encourage weed invasion, and alter natural flow regimes. Native species often rely on such flow regimes for spawning, juvenile development, and migration.

In some areas, especially along the U.S. coast, polluted runoff from roads and highways may be the largest source of water pollution. For example, about 75 percent of the toxic chemicals getting to Seattle, Washington's Puget Sound are carried by stormwater that runs off paved roads and driveways, rooftops, yards, and other developed land.

Urban Flooding

Stormwater is a major cause of urban flooding. Urban flooding is the inundation of land or property in a built-up environment caused by stormwater overwhelming the capacity of drainage systems, such as storm sewers. Although triggered by single events such as flash flooding or snow melt, urban flooding is a condition, characterized by its repetitive, costly and systemic impacts on communities. In areas susceptible to urban flooding, backwater valves and other infrastructure may be installed to mitigate losses.

Where properties are built with basements, urban flooding is the primary cause of basement and sewer backups. Although the number of casualties from urban flooding is usually limited, the economic, social and environmental consequences can be considerable: in addition to direct damage to property and infrastructure (highways, utilities and services), chronically wet houses are linked to an increase in respiratory problems and other illnesses.

Urban flooding has significant economic implications. In the U.S., industry experts estimate that wet basements can lower property values by 10 to 25 percent and are cited among the top reasons for not purchasing a home. According to the Federal Emergency Management Agency almost 40 percent of small businesses never reopen their doors following a flooding disaster. In the UK, urban flooding is estimated to cost £270 million a year (as of 2007) in England and Wales; 80,000 homes are at risk.

A study of Cook County, Illinois, identified 177,000 property damage insurance claims made across 96 percent of the county's ZIP codes over a five-year period from 2007 to 2011. This is the equivalent of one in six properties in the County making a claim. Average payouts per claim were $3,733 across all types of claims, with total claims amounting to $660 million over the five years examined.

An example of Urban flooding control project is the Brays Bayou Greenway Framework. The Brays Bayou Greenway Framework is federally funded improvement project. Brays Bayou starts from the Bayou's mouth at Buffalo Bayou and the Ship Channel in the east to the Barker Reservoir and

George Bush Park in the west. In aim to identify a broad set of potential recreation and open space opportunities along the 35 miles of Brays Bayou, Brays Bayou and its tributaries provide drainage to a watershed of approximately 88,000 acres south of downtown Houston, Texas. Project Brays responses to the present flooding problems in Houston and creates a short-term solution by improving the bayou's drainage capacity, also considered a long-term solution for flooding preventing.

Stormwater Management

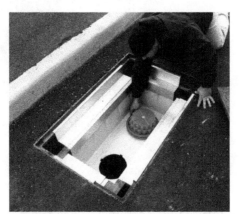

Stormwater filtration system for urban runoff

Managing the quantity and quality of stormwater is termed, "Stormwater Management." The term *Best Management Practice* (BMP) or stormwater control measure (SCM) is often used to refer to both structural or engineered control devices and systems (e.g. retention ponds) to treat or store polluted stormwater, as well as operational or procedural practices (e.g. street sweeping). Stormwater management includes both technical and institutional aspects, including:

- control of flooding and erosion;

- control of hazardous materials to prevent release of pollutants into the environment (source control);

- planning and construction of stormwater systems so contaminants are removed before they pollute surface waters or groundwater resources;

- acquisition and protection of natural waterways or rehabilitation;

- building "soft" structures such as ponds, swales, wetlands or green infrastructure solutions to work with existing or "hard" drainage structures, such as pipes and concrete channels;

- development of funding approaches to stormwater programs potentially including stormwater user fees and the creation of a stormwater utility;

- development of long-term asset management programs to repair and replace aging infrastructure;

- revision of current stormwater regulations to address comprehensive stormwater needs;

- enhancement and enforcement of existing ordinances to make sure property owners consider the effects of stormwater before, during and after development of their land;

- education of a community about how its actions affect water quality, and about what it can do to improve water quality; and

- planning carefully to create solutions before problems become too great.

Integrated Water Management

Rain garden designed to treat stormwater from adjacent parking lot

Integrated water management (IWM) of stormwater has the potential to address many of the issues affecting the health of waterways and water supply challenges facing the modern urban city.

Also known as low impact development (LID) in the United States, or Water Sensitive Urban Design (WSUD) in Australia, IWM has the potential to improve runoff quality, reduce the risk and impact of flooding and deliver an additional water resource to augment potable supply.

The development of the modern city often results in increased demands for water supply due to population growth, while at the same time altered runoff predicted by climate change has the potential to increase the volume of stormwater that can contribute to drainage and flooding problems. IWM offers several techniques including stormwater harvest (to reduce the amount of water that can cause flooding), infiltration (to restore the natural recharge of groundwater), biofiltration or bioretention (e.g., rain gardens) to store and treat runoff and release it at a controlled rate to reduce impact on streams and wetland treatments (to store and control runoff rates and provide habitat in urban areas).

There are many ways of achieving LID. The most popular is to incorporate land-based solutions to reduce stormwater runoff through the use of retention ponds, bioswales, infiltration trenches, sustainable pavements (such as permeable paving), and others noted above. LID can also be achieved by utilizing engineered, manufactured products to achieve similar, or potentially better, results as land-based systems (underground storage tanks, stormwater treatment systems, biofilters, etc.). The proper LID solution is one that balances the desired results (controlling runoff and pollution) with the associated costs (loss of usable land for land-based systems versus capital cost of manufactured solution). Green (vegetated) roofs are also another low cost solution.

IWM as a movement can be regarded as being in its infancy and brings together elements of drainage science, ecology and a realization that traditional drainage solutions transfer problems further downstream to the detriment of our environment and precious water resources.

Regulations

Federal Requirements

Retention basin for management of stormwater

In the United States, the Environmental Protection Agency (EPA) is charged with regulating stormwater pursuant to the Clean Water Act (CWA). The goal of the CWA is to restore all "Waters of the United States" to their "fishable" and "swimmable" conditions. Point source discharges, which originate mostly from municipal wastewater (sewage) and industrial wastewater discharges, have been regulated since enactment of the CWA in 1972. Pollutant loadings from these sources are tightly controlled through the issuance of National Pollution Discharge Elimination System (NPDES) permits. However, despite these controls, thousands of water bodies in the U.S. remain classified as "impaired," meaning that they contain pollutants at levels higher than is considered safe by EPA for the intended beneficial uses of the water. Much of this impairment is due to polluted runoff.

To address the nationwide problem of stormwater pollution, Congress broadened the CWA definition of "point source" in 1987 to include industrial stormwater discharges and municipal separate storm sewer systems ("MS4"). These facilities were required to obtain NPDES permits.

State and Local Requirements

A silt fence, a type of sediment control, installed on a construction site

EPA has authorized 46 states to issue NPDES permits. In addition to implementing the NPDES requirements, many states and local governments have enacted their own stormwater management laws and ordinances, and some have published stormwater treatment design manuals. Some of these state and local requirements have expanded coverage beyond the federal requirements. For example, the State of Maryland requires erosion and sediment controls on construction sites of 5,000 sq ft (460 m²) or more. It is not uncommon for state agencies to revise their requirements and impose them upon counties and cities; daily fines ranging as high as $25,000 can be imposed for failure to modify their local stormwater permitting for construction sites, for instance.

Nonpoint Source Pollution Management

Agricultural runoff (except for concentrated animal feeding operations, or "CAFO") is classified as nonpoint source pollution under the CWA. It is not included in the CWA definition of "point source" and therefore not subject to NPDES permit requirements. The 1987 CWA amendments established a non-regulatory program at EPA for nonpoint source pollution management consisting of research and demonstration projects. Related programs are conducted by the Natural Resources Conservation Service (NRCS) in the U.S. Department of Agriculture.

Public Education Campaigns

Education is a key component of stormwater management. A number of agencies and organizations have launched campaigns to teach the public about stormwater pollution, and how they can contribute to solving it. Thousands of local governments in the U.S. have developed education programs as required by their NPDES stormwater permits.

One example of a local educational program is that of the West Michigan Environmental Action Council (WMEAC), which has coined the term *Hydrofilth* to describe stormwater pollution, as part of its "15 to the River" campaign. (During a rain storm, it may take only 15 minutes for contaminated runoff in Grand Rapids, Michigan to reach the Grand River.) Its outreach activities include a rain barrel distribution program and materials for homeowners on installing rain gardens.

Other public education campaigns highlight the importance of green infrastructure in slowing down and treating stormwater runoff. DuPage County Stormwater Management launched the "Love Blue. Live Green." outreach campaign on social media sites to educate the public on green infrastructure and other best management practices for stormwater runoff. Articles, websites, pictures, videos and other media are disseminated to the public through this campaign.

History

Since humans began living in concentrated village or urban settings, stormwater runoff has been an issue. During the Bronze Age, housing took a more concentrated form, and impervious surfaces emerged as a factor in the design of early human settlements. Some of the early incorporation of stormwater engineering is evidenced in Ancient Greece.

A specific example of an early stormwater runoff system design is found in the archaeological recovery at Minoan Phaistos on Crete.

Stormwater Harvesting

Stormwater harvesting is the collection, accumulation, treatment or purification, and storing of stormwater for its eventual reuse. It differs from rainwater harvesting as the runoff is collected from drains or creeks, rather than roofs. It can also include other catchment areas from man made surfaces, such as roads, or other urban environments such as parks, gardens and playing fields. Water that comes in contact with impervious surfaces becomes polluted and is denominated surface runoff. As the water travels more distance over impervious surfaces it collects an increasing amount of pollutants.

The main challenge stormwater harvesting poses is the removal of pollutants in order to make this water available for reuse.

Ground Catchment Systems

Catching hillside run-off water

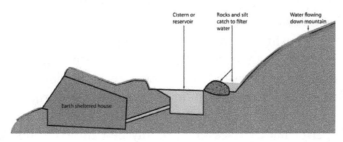

Ground catchments systems channel water from a prepared catchment area into storage. Generally they are only considered in areas where rainwater is very scarce and other sources of water are not available. They are more suited to small communities than individual families. If properly designed, ground catchment systems can collect large quantities of rainwater.

References

- Debo, Tom; Reese, Andrew (2003). "Chapter 2. Stormwater Management Programs". Municipal Stormwater Management. Boca Raton, FL: CRC Press. ISBN 1-56670-584-3.
- "Reading: Urban Stormwater Management in the United States | The National Academies Press". www.nap.edu. Retrieved 2015-09-19.

5

Water Integration System

Integrated water resources management is defined as a process which promotes the coordinated development and management of water, land and related resources. The program takes into account both urban and rural areas for water resources development. The content gives an overview of the subject matter incorporating all the major aspects of water integration

Integrated Urban Water Management

Integrated urban water management (IUWM) is a philosophy of varying definitions and interpretations. According to the authors of the book entitled, "Integrated Urban Water Management: Humid Tropics", IUWM is described as the practice of managing freshwater, wastewater, and storm water as components of a basin-wide management plan. It builds on existing water supply and sanitation considerations within an urban settlement by incorporating urban water management within the scope of the entire river basin. One of the early champions of IUWM, SWITCH is a research program funded by the European Union and seeks to shift urban water management away from ad hoc solutions to a more integrated approach. IUWM within an urban water system can also be conducted by performance assessment of any new intervention strategies by developing a holistic approach which encompasses various system elements and criteria including sustainability type ones in which integration of water system components including water supply, waste water and storm water subsystems would be advantageous. Simulation of metabolism type flows in urban water system can also be useful for analysing processes in urban water cycle of IUWM.

IUWM is commonly seen as a strategy for achieving the goals of Water Sensitive Urban Design. IUWM seeks to change the impact of urban development on the natural water cycle, based on the premise that by managing the urban water cycle as a whole; a more efficient use of resources can be achieved providing not only economic benefits but also improved social and environmental outcomes. One approach is to establish an inner, urban, water cycle loop through the implementation of reuse strategies. Developing this urban water cycle loop requires an understanding both of the natural, pre-development, water balance and the post-development water balance. Accounting for flows in the pre- and post-development systems is an important step toward limiting urban impacts on the natural water cycle.

Components

Activities under the IUWM include the following:

- Improve water supply and consumption efficiency

- Upgrade drinking water quality and wastewater treatment

- Increase economic efficiency of services to sustain operations and investments for water, wastewater, and stormwater management

- Utilize alternative water sources, including rainwater, and reclaimed and treated water

- Engage communities to reflect their needs and knowledge for water management

- Establish and implement policies and strategies to facilitate the above activities

- Support capacity development of personnel and institutions that are engaged in IUWM

According to Australia's Commonwealth Scientific and Industrial Research Organisation (CSIRO), IUWM requires the management of the urban water cycle in coordination with the hydrological water cycle which are significantly altered by urban landscapes and its correlation to increasing demand. Under natural conditions the water inputs at any point in the system are precipitation and overland flows; while the outputs are via surface flows, evapo-transpiration and groundwater recharge. The large volumes of piped water introduced with the change to an urban setting and the introduction of vast impervious areas strongly impact the water balance, increasing in-flows and dramatically altering the out-flow components.

Approaches

- The Agenda 21 (UN Department for Sustainable Development, 1992) has worked out the Dublin Principles for Integrated water resources management in more detail for urban areas. One of the objectives of Agenda 21 is to develop environmentally sound management of water resources for urban use.

- The Bellagio Statement formulated by the Environmental Sanitation Working Group of the Water Supply and Sanitation Collaborative Council in 2000 include principals such as: Human dignity, quality of life, environmental security, an open stakeholder process, and many others.

- The UNEP 3 Step Strategic Approach developed in 2005 is based on the application of the "Cleaner Production approach" that has been successful in the industrial sector. The three steps are: Prevention, Treatment for reuse, and Planned discharge with stimulation of self-purification capacity.

- UNESCO's Institute for Water Education seeks to build on the progress made by the Bellagio Statement and UNEP's 3-step approach by developing the *SWITCH* approach to IUWM. Components include: the addition of a sustainability assessment, new methods of planning urban water systems, and modifications to planning and strategy development.

Examples

An example of IUWM is the Catskill/ Delaware water system that provides 1.4 billion US gallons (5,300,000 m³) of water per day, including to all of New York City. The IUWM process included an extensive stakeholder engagement process, whereby the needs of all parties were included into the final management plan. A partnership was created between New York City, the agricultural community, and the federal government. The case has become a model for successful IUWM.

Grey Water Systems

Grey water (also written; greywater, gray water, or graywater) is water used with appliances that do not involve or encounter human waste. It gets its name relative to black water which is heavily contaminated with human waste. Different resources suggest what equipment produce grey or black water. However, it is most commonly accepted that bathtubs, showers, washbasins, washing machines, and laundry tubs produce grey water, whereas toilets, sinks, and dishwashers are classified as black water sources.

In 1989 Santa Barbara became the first district in the United States to legalize the recycling and reuse of grey water. Since then, grey water has become a part of integrated urban water management. It addresses the practice of managing wastewater at the residential scale. The premise of grey water reuse is the concept that the average household uses most of its water indoors (roughly 60%), meaning much of that water can be reused to provide for the water required to support irrigation. Additionally, most domestic appliances automatically collect grey water in order for it to be disposed.

There are three types of grey water systems each of which has different requirements, codes, and sizing specifications. However, they share standards to meet health and safety regulations:

The Arizona R18-9-711. Type 1 Reclaimed Water General Permit for Gray Water, stands as the model for grey water system regulation

1. Human contact with gray water and soil irrigated by gray water is avoided;

2. Gray water originating from the residence is used and contained within the property boundary for household gardening, composting, lawn watering, or landscape irrigation;

3. Surface application of gray water is not used for irrigation of food plants, except for citrus and nut trees;

4. The gray water does not contain hazardous chemicals derived from activities such as cleaning car parts, washing greasy or oily rags, or disposing of waste solutions from home photo labs or similar hobbyist or home occupational activities;

5. The application of gray water is managed to minimize standing water on the surface;

6. The gray water system is constructed so that if blockage, plugging, or backup of the system occurs, gray water can be directed into the sewage collection system or onsite wastewater treatment and disposal system, as applicable. The gray water system may include a means of filtration to reduce plugging and extend system lifetime;

7. Any gray water storage tank is covered to restrict access and to eliminate habitat for mosquitoes or other vectors;

8. The gray water system is sited outside of a floodway;

9. The gray water system is operated to maintain a minimum vertical separation distance of at least five feet from the point of gray water application to the top of the seasonally high groundwater table;

10. For residences using an onsite wastewater treatment facility for black water treatment and disposal, the use of a gray water system does not change the design, capacity, or reserve area requirements for the onsite wastewater treatment facility at the residence, and ensures that the facility can handle the combined black water and gray water flow if the gray water system fails or is not fully used;

11. Any pressure piping used in a gray water system that may be susceptible to cross connection with a potable water system clearly indicates that the piping does not carry potable water;

12. Gray water applied by surface irrigation does not contain water used to wash diapers or similarly soiled or infectious garments unless the gray water is disinfected before irrigation; and

13. Surface irrigation by gray water is only by flood or drip irrigation.

A complex grey water system provides for a development with a substantial discharge (greater than 250 gallons) per day. It requires a written construction permit submitted to the enforcing agency.

A simple grey water system is sized to serve a one or two family home with a medium level water discharge (maximum 250 gallons) per day. It too, requires a written construction permit submitted to the Enforcing Agency.

A clothes washer grey water system is sized to recycle the grey water of a one or two family home using the reclaimed water of a washing machine (produces 15 gallons per person per day). It relies on either the pump from the washing machine or gravity to irrigate. This particular system is the most common and least restricted system. In most states, this system does not require construction permits. This system is often characterized as Laundry to Landscape (L2L). The system relies on valves, draining to a mulch basin, or the area of irrigation for certain landscape features (a mulch basin for a tree requires 12.6 ft²). The drip system must be calibrated to avoid uneven distribution of grey water or overloading.

Delivery Methods

- Above Grade Gravity-Fed Irrigation

Often used for washing machine or tub with gravity-fed irrgation systems

1. Grey water leaves source and passes through three-way diversion valve
2. Water enters either gully trap or surge tank
3. Tank water enters filter bag
4. Filtered water falls to irrigation system

- Submersible Pump Diversion to Irrigation

Often used for low level of contamination in water, providing that unsuitable water enters sewer

1. Grey Water discharges over a screen filter which can be diverted to a gully with a valve leading to the sewer

2. Screen filtered water enters a submersed pump

3. Pump sends water to irrigation system

- Above Grade Gravity-Fed Pumped Irrigation

Often used for laundry or bathroom grey water systems with pumped irrigation

1. Grey water leaves through three-way diversion valve

2. Enters surge tank

3. Tank water enters filter bag

4. Filtered water enters submersible pump

5. Pump supplies water to irrigation system

Indoor Reuse

Recycled grey water from domestic appliances can be also be used to flush toilet. It application is based on standards set by plumbing codes. Indoor grey water reuse requires an efficient cleaning tank for insoluble waste, as well as a well regulated control mechanism.

The Uniform Plumbing Code, adopted in some U.S. jurisdictions, prohibits greywater use indoors.

Construction of Grey Water Systems

A written construction permit is required prior to building, fixing, relocating, or altering any system that *requires* a permit. Permits require usage data, member sizing, and soil conditions (to reduce pooling of grey water). Different states and regions are subject to different requirements of plans and system specifications. Standards set by state dictate codes for grey water systems based on definition of grey water, system type, and permit requirement. In many cases, construction can be done without contracting a professional.

Testing and Inspection

A grey water system that requires a permit must accurately determine the absorptive capacity of soils. Certain tests may be required for irregular soils, or undocumented areas.

A permitted system is required to be inspected. During construction the system is to be left uncovered. Additionally, implementation of certain devices can be prescribed in codes and requirements. A Reduced Pressure (RP) backflow device at property entry makes it easier to assess any amounts of crossover between grey water and a potentially potable water supply.

Challenges

The biggest concern in Integrated Urban Water Management systems, specifically grey water reuse is the introduction of untreated water into a landscape. It is important that grey water systems maximize natural purification through healthy topsoil, avoiding contact between greywater before and after filtration. Untreated grey water has the capacity to meet runoff and ultimately pollute water systems.

Research suggests that evaporated grey water can leave microorganisms that can be harmful if breathed or consumed. It is best practice to not use greywater in a sprinkler system, for this reason. Direct application of grey water can leave the aforementioned microorganisms on foliage. Grey water should not be used on fruits or vegetables (unless applied very carefully and specifically to the roots – although most state codes won't allow this).

Excess grey water which may not percolate into the soil could become runoff often leading the untreated water to waterways. Grey water should not be applied to saturated soil, and should be used conservatively.

Many household cleaners contain ingredients which can't be removed by a typical grey water filtration system. This requires an assessment of which substances will alter the quality of the grey water. Additionally, not all household appliances should be used for grey water recycling. Some should be diverted separately to sewage.

Grey water filtration systems are not equipped to handle the highest levels contaminants. Discretion is required as to when to use the grey water system. This includes consideration of what appliances to connect, as well as how much water is being processed at any given time.

One of the most significant challenges for IUWM could be securing a consensus on the definition of IUWM and the implementation of stated objectives at operational stages of projects. In the developing world there is still a significant fraction of the population that has no access to proper water supply and sanitation. At the same time, population growth, urbanization and industrialization continue to cause pollution and depletion of water sources. In the developed world, pollution of water sources is threatening the sustainability of urban water systems. Climate change is likely to affect all urban centers, either with increasingly heavy storms or with prolonged droughts, or perhaps both. To address the challenges facing IUWM it is crucial to develop good approaches, so that policy development and planning are directed towards addressing these global change pressures, and to achieving truly sustainable urban water systems.

Integrated Water Resources Management

Integrated water resources management (IWRM) has been defined by the Global Water Partnership (GWP) as "a process which promotes the coordinated development and management of water, land and related resources, in order to maximize the resultant economic and social welfare in an equitable manner without compromising the sustainability of vital ecosystems".

Development

The development of IWRM was particularly recommended in the final statement of the ministers at the International Conference on Water and the Environment in 1992 (so called the Dublin principles). This concept aims to promote changes in practices which are considered fundamental to improved water resource management. In the current definition, IWRM rests upon three principles that together act as the overall framework:

1. Social equity: ensuring equal access for all users (particularly marginalised and poorer user groups) to an adequate quantity and quality of water necessary to sustain human well-being.

2. Economic efficiency: bringing the greatest benefit to the greatest number of users possible with the available financial and water resources.

3. Ecological sustainability: requiring that aquatic ecosystems are acknowledged as users and that adequate allocation is made to sustain their natural functioning.

IWRM practices depend on context; at the operational level, the challenge is to translate the agreed principles into concrete action.

Implementation

Some of the cross-cutting conditions that are also important to consider when implementing IWRM are:

- Political will and commitment

- Capacity development

- Adequate investment, financial stability and sustainable cost recovery

- Monitoring and evaluation

IWRM should be viewed as a process rather than a one-shot approach - one that is long-term and iterative rather than linear in nature. As a process which seeks to shift water development and management systems from their currently unsustainable forms, IWRM has no fixed beginnings or endings.

Furthermore, there is not one correct administrative model. The art of IWRM lies in selecting, adjusting and applying the right mix of these tools for a given situation.

References

- Behzadian, K; Kapelan, Z (2015). "Advantages of integrated and sustainability based assessment for metabolism based strategic planning of urban water systems". Science of The Total Environment. 527-528: 220–231. doi:10.1016/j.scitotenv.2015.04.097.

- Behzadian, k; Kapelan, Z (2015). "Modelling metabolism based performance of an urban water system using WaterMet2". Resources, Conservation and Recycling. 99: 84–99. doi:10.1016/j.resconrec.2015.03.015.

6

Irrigation Strategies and Concepts

The major concepts of irrigation are discussed in this chapter. Deficit irrigation, flexible barge, hippo water roller and peak water are important topics related to irrigation strategies. The chapter strategically encompasses and incorporates the major components and key concepts of irrigation strategies, providing a complete understanding.

Deficit Irrigation

Deficit irrigation (DI) is a watering strategy that can be applied by different types of irrigation application methods. The correct application of DI requires thorough understanding of the yield response to water (crop sensitivity to drought stress) and of the economic impact of reductions in harvest. In regions where water resources are restrictive it can be more profitable for a farmer to maximize crop water productivity instead of maximizing the harvest per unit land. The saved water can be used for other purposes or to irrigate extra units of land. DI is sometimes referred to as incomplete supplemental irrigation or regulated DI.

Definition

Deficit irrigation (DI) has been reviewed and defined as follows:

"Deficit irrigation is an optimization strategy in which irrigation is applied during drought-sensitive growth stages of a crop. Outside these periods, irrigation is limited or even unnecessary if rainfall provides a minimum supply of water. Water restriction is limited to drought-tolerant phenological stages, often the vegetative stages and the late ripening period. Total irrigation application is therefore not proportional to irrigation requirements throughout the crop cycle. While this inevitably results in plant drought stress and consequently in production loss, DI maximizes irrigation water productivity, which is the main limiting factor (English, 1990). In other words, DI aims at stabilizing yields and at obtaining maximum crop water productivity rather than maximum yields (Zhang and Oweis, 1999)."

Crop Water Productivity

Crop water productivity (WP) or water use efficiency (WUE) expressed in kg/m³ is an efficiency term, expressing the amount of marketable product (e.g. kilograms of grain) in relation to the amount of input needed to produce that output (cubic meters of water). The water used for crop production is referred to as crop evapotranspiration. This is a combination of water lost by evaporation from the soil surface and transpiration by the plant, occurring simultaneously. Except by modeling, distinguishing between the two processes is difficult. Representative values of WUE for cereals at field level, expressed with evapotranspiration in the denominator, can vary between 0.10 and 4 kg/m3.

Experiences with Deficit Irrigation

For certain crops, experiments confirm that DI can increase water use efficiency without severe yield reductions. For example for winter wheat in Turkey, planned DI increased yields by 65% as compared to winter wheat under rainfed cultivation, and had double the water use efficiency as compared to rainfed and fully irrigated winter wheat. Similar positive results have been described for cotton. Experiments in Turkey and India indicated that the irrigation water use for cotton could be reduced to up to 60 percent of the total crop water requirement with limited yield losses. In this way, high water productivity and a better nutrient-water balance was obtained.

Certain Underutilized and horticultural crops also respond favorably to DI, such as tested at experimental and farmer level for the crop quinoa. Yields could be stabilized at around 1.6 tons per hectare by supplementing irrigation water if rainwater was lacking during the plant establishment and reproductive stages. Applying irrigation water throughout the whole season (full irrigation) reduced the water productivity. Also in viticulture and fruit tree cultivation, DI is practiced.

Scientists affiliated with the Agricultural Research Service (ARS) of the USDA found that conserving water by forcing drought (or deficit irrigation) on peanut plants early in the growing season has shown to cause early maturation of the plant yet still maintain sufficient yield of the crop. Inducing drought through deficit irrigation earlier in the season caused the peanut plants to physiologically "learn" how to adapt to a stressful drought environment, making the plants better able to cope with drought that commonly occurs later in the growing season. Deficit irrigation is beneficial for the farmers because it reduces the cost of water and prevents a loss of crop yield (for certain crops) later on in the growing season due to drought. In addition to these findings, ARS scientists suggest that deficit irrigation accompanied with conservation tillage would greatly reduce the peanut crop water requirement.

For other crops, the application of deficit irrigation will result in a lower water use efficiency and yield. This is the case when crops are sensitive to drought stress throughout the complete season, such as maize.

Apart from university research groups and farmers associations, international organizations such as FAO, ICARDA, IWMI and the CGIAR Challenge Program on Water and Food are studying DI.

Reasons for Increased Water Productivity Under Deficit Irrigation

If crops have certain phenological phases in which they are tolerant to water stress, DI can increase the ratio of yield over crop water consumption (evapotranspiration) by either reducing the water loss by unproductive evaporation, and/or by increasing the proportion of marketable yield to the totally produced biomass (harvest index), and/or by increasing the proportion of total biomass production to transpiration due to hardening of the crop - although this effect is very limited due to the conservative relation between biomass production and crop transpiration, - and/or due to adequate fertilizer application and/or by avoiding bad agronomic conditions during crop growth, such as water logging in the root zone, pests and diseases, etc.

Advantages

The correct application of DI for a certain crop:

- maximizes the productivity of water, generally with adequate harvest quality;

- allows economic planning and stable income due to a stabilization of the harvest in comparison with rainfed cultivation;

- decreases the risk of certain diseases linked to high humidity (e.g. fungi) in comparison with full irrigation;

- reduces nutrient loss by leaching of the root zone, which results in better groundwater quality and lower fertilizer needs as for cultivation under full irrigation;

- improves control over the sowing date and length of the growing period independent from the onset of the rainy season and therefore improves agricultural planning.

Constraints

A number of constraints apply to deficit irrigation:

- Exact knowledge of the crop response to water stress is imperative.

- There should be sufficient flexibility in access to water during periods of high demand (drought sensitive stages of a crop).

- A minimum quantity of water should be guaranteed for the crop, below which DI has no significant beneficial effect.

- An individual farmer should consider the benefit for the total water users community (extra land can be irrigated with the saved water), when he faces a below-maximum yield;

- Because irrigation is applied more efficiently, the risk for soil salinization is higher under DI as compared to full irrigation.

Modeling

Field experimentation is necessary for correct application of DI for a particular crop in a particular region. In addition, simulation of the soil water balance and related crop growth (crop water productivity modeling) can be a valuable decision support tool. By conjunctively simulating the effects of different influencing factors (climate, soil, management, crop characteristics) on crop production, models allow to (1) better understand the mechanism behind improved water use efficiency, to (2) schedule the necessary irrigation applications during the drought sensitive crop growth stages, considering the possible variability in climate, to (3) test DI strategies of specific crops in new regions, and to (4) investigate the effects of future climate scenarios or scenarios of altered management practices on crop production.

Flexible Barge

A flexible barge is a fabric barge (non-rigid) for the transportation of bulk fresh water or other liquid bulk items such as chemicals or oil.

Flexible barge air test

History

One such barge is called the Dracone Barge invented in 1956. Other similar devices are the Spragg Bag invented in the 1980s, the water bag proposed by Nordic Water Supply in the late 1990s and the more recent REFRESH modular waterbag, developed in the 2010s. Terry Spragg of Manhattan Beach, California, builds flexible fabric barges for the transportation of bulk fresh water and is the reason why his product is referred to as the "Spragg Bag." In the 1970s Spragg was a promoter of icebergs as a large source of fresh water, but soon realized this was impractical. He then put his skills into developing the waterbag technology starting in the 1980s. Spragg has worked on and perfected this over the last twenty years with his associates. The first field test of his waterbag was in December 1990. The waterbag was 75 meters (246 feet) long and it contained approximately 3,000 cubic metres (790,000 US gal) of fresh water. It was towed from the Port Angeles harbor in the state of Washington. Another test was done in 1996 with a 100-mile (160 km) voyage from Port Angeles to Seattle, Washington. This test ended on April 29 when the fabric of one of the two bags under tow developed a tear. Spragg says that his next goal is to run another test voyage demonstration between Northern and Southern California and a demonstration of the waterbag technology in the Middle East as well as around the world. There are various reasons why it has been difficult to gain support for demonstrating the viability of waterbag technology in California and around the world. Spragg claims when two waterbags pass underneath the Golden Gate Bridge for the first time in history the media will let the whole world know about it. A novel, *Water, War, and Peace*, has been completed that details the solutions waterbag technology offers to the complex political problems surrounding water issues throughout the Middle East, the United States, and the world.

Nordic Water Supply waterbag

The Norwegian company Nordic Water Supply developed a 10.800 m³ bag in 1997 under an agreement with the Turkish government to transport freshwater to Northern Cyprus. Within two years at least 7 million m³ of water had to be delivered annually at a cost of €2.7M per year, with volumes growing over time, but the actual transport only amounted to 4 million m³ in four years and the contract was discontinued by Turkish authorities. As a result, NWS went out of business and was de-listed from the Oslo Stock Exchange in 2003. NSW's waterbag technology was acquired by the Monohakobi Institute of Technology in Japan.

REFRESH waterbag

The REFRESH waterbag was developed by a consortium of companies and research institutes from Greece, Spain, Italy, Turkey and the Czech Republic within two European FP7 projects, RE-FRESH (running from 2010 to 2012) and the follow-up XXL-REFRESH (running from 2013 to 2015). The first project was focused on validation of the concept of modular waterbag; it developed a small scale prototype of 200 m³ capacity, tested in Greece in 2012. The second project was focused on scale-up and partial redesign of the REFRESH system. At the end of the second project the REFRESH waterbag concept reached commercial scale and a 2500 m³ system made of five 500m³ modules was tested offshore Spain in 2015. The length of the waterbag was 60 m long.

The REFRESH concept is different from concepts of waterbag proposed earlier, based on huge monolithic containers (as the one proposed by Nordic Water Supply) or "trains" of smaller containers each one being sealed in itself (as the Spragg bag). The REFRESH waterbag is made of a series of modules, each one being a cylinder open at both bases, joined by watertight zippers. This makes it possible to perform all "dry" operations on ground at the level of single modules, overcoming the handling problems of monolithic containers and improves the behaviour in navigation with respect to the "trains" of connected bags.

Technology

Flexible barge filled with fresh water

Flexible barge zipper

The 1995 associated Spragg patents (#5,413,065 and #5,488,921) indicate that the inventions relate to a flexible fabric barge technology or combination of several barges made of a rubber polyurethane material. The main body portion of a flexible fabric barge is cylindrical in shape. The barge can be used by itself or as several connected flexible fabric barges that can be towed through the open ocean under extreme conditions. The patents further explain that the goal of Spragg's inventions are a practical water delivery system of fresh drinkable water that could be delivered to dry regions worldwide that have a shortage of potable water.

One of the flexible fabric barge concepts aims at an economical delivery system for fresh water that would be considerably cheaper than desalination plants, rigid ships, tanker trucks, conventional barges, aqueducts or pipeline transport. Waterbags are more economical and better for the environment than desalination of the seas and oceans.

The flexible fabric barge is a plastic container that is buoyant. It can be linked together with an attachment system to other flexible fabric barges to make a "train." The "waterbags", which are sometimes referred to as balloons or bladders, are made up of a vinyl lining inside a stronger material net. It is a type of "fabric water pipeline" when several are strung together for transport of liquids by tugboat through seas or open oceans to remote locations.

Zipper

Zippers play an important part in extending the capacity of the waterbag beyond what is practically achievable by a single textile piece. The Spragg and REFRESH concepts both feature prominently zippers, albeit with a fundamental difference in its function.

In the Spragg design, large waterbags are connected together like boxcars in a "train" fashion to increase the amount of liquids delivered at a time. It is estimated that the flexible barges could be as large as 14 meters in diameter and 150 meters in length, holding up to 17,000 cubic meters of fresh water or any other transportable liquid. Theoretically as many as 50 to 60 "waterbags" could be connected to one another and towed, although such a test has not been done to date.

REFRESH waterbag's fabric with embedded zipper

Engineers suggest that the fabric barge can hold a "train" of 4,500,000 US gallons (17,000 m³) of liquids.

A zipper coupler system is 10 times stronger than what is pulled by a 4,300 hp tug. Laborde Marine estimates a 4,300 horsepower tug with a bollard pull of 110,000 pounds can pull a "train" of fifty or so flexible barges weighing up to 1,300,000 tons. The "train" speed would be about 3 knots. This is over 700 acre feet (8,600,000 hl) of fresh water or other liquid per trip.

In the REFRESH design, the container itself is assembled on shore starting from planar cuts of fabric. Zippers run all along the perimeter of the fabric and make it possible to join an indefinite number of modules. Since each module is not closed by itself, zippers need to be watertight in order to ensure that no seawater leaks in.

Invention

Spragg Flexible barges coupled with zippers

The basic invention is a device for the delivery of huge amounts of fresh water (700,000 to 4,500,000 US gallons (17,000 m³) in each bag) at one time in a hostile wind and wave environment typical of oceans and large seas. Since freshwater is lighter than seawater the filled "bladders" (as they are sometimes referred to) float on top, similarly to icebergs, with little above the surface and most below the surface. Fresh water can be taken from rivers just before it discharges into salty seas or oceans, which then would not interfere with salmon spawning.

The associated coupler and zipper patent describes that to be economically feasible there should be several such flexible barges towed at one time. The greater the volume of water that can be delivered per trip, the better the economics. This string of barges would typically consist of barges in size from 25 to 50 feet (15 m) in diameter and 200 to 800 feet (240 m) in length each.

The unique characteristic of the Spragg Bag system is not the large volume of water in each bag, but what is called the world's strongest zipper (produced by Italian company Ziplast) that allows connection of several bags together in long trains. The large connecting zipper can be operated manually or by remote control with radio signals. The string of such flexible fabric "waterbags" may be coupled to a barge via a reinforced fabric nose cone where a tow line is attached.

The REFRESH scheme is enabled by a specialty zipper, again developed by Ziplast, that uses a completely different tooth engagement design able to keep the strength of the original "Spragg" zipper while adding watertightness. Tests performed by the Spanish research centre AIMPLAS have confirmed that the zipper is able to stay watertight even when in tension.

Each "waterbag" is generally filled to 80-90% capacity (so that it is not stiff and remains able to adapt to deformations when turning) and is flat across its top.

Economics

500,000 acre feet (620,000,000 m³) per year waterbag loading system designed by CH2M-HILL

One San Francisco area reporter writes that waterbag technology would provide economical fresh water delivery to the Monterey Peninsula district and a solution to the shortage of fresh water in the area. He reports that the average family of four uses one acre-foot of water a year. This costs over $1000 for delivery using conventional methods, however this same amount of water delivered by Spragg Bags would cost about 30% less. Another newspaper reporter explains that towing Spragg Waterbags is environmentally friendly and is more economical than carrying water in ships or water tanker trucks or even using conventional rigid pipelines. An article in the July issue of ECONOMIST Magazine in 2008 explains that worldwide there is enough water for all, however most is often in the wrong place at the wrong time and it is just too expensive to transport.

Waterbag technology offers an easy and inexpensive solution to the problem of today's expensive conventional water transportation. It eliminates the difficulty of transporting water long distances by using the ocean as the means of transport. Waterbags considerably lower the capital costs and operating costs of moving fresh water from place to place. If a train was able to only move one or two box cars at a time, rather than in a train of dozens of boxcars at a time, it would not be very efficient and extremely expensive. Linking waterbags into "trains" of strings of waterbags and moving them through the ocean increases the economics of water transportation making it a viable practical option.

The cost to transport water 300 to 800 miles (1,300 km) through the ocean, based on deliveries of 5 million US gallons per day (19,000 m³/d) to 10 million US gallons per day (38,000 m³/d), is estimated to be between $350 to $450 per acre foot, depending on the length of the voyage and the amount of water delivered per trip. Increasing the amount of water delivered per day in each waterbag train will help to significantly reduce the cost of the water delivered. Once the reliability of the waterbag delivery system has proven its economics and reliability it will just be a matter of adding more waterbags to the trains, and more trains to the system in order to increase the amount of water delivered to selected locations, while also reducing the cost of the water delivered. Based on the increasing reliability of the waterbag delivery system over time, it should be possible to be able to economically deliver 100,000's of acre feet per year to many coastal locations around the world.

According to the inventor of the Spragg bag, the total cost of delivering fresh water down the California coast by his waterbag technology for a distance of 800 miles (1,300 km) from British Columbia to Monterey would cost about $966 per acre-foot per year. Keith Spain in a study for a Master Of Arts then shows in an analysis that it would save the residents of the Monterey Peninsula some $1,134 per-acre foot otherwise using a desalination plant. This is a savings of over $19 million per year for the Monterey taxpayers. This number assumes a usage of approximately 17,000 acre feet (21,000,000 m³) per year (17,000 X $1,134 = $19,278,000 savings).

Applications

Flexible barge with camera crew on top filming an actual application.

One application seen is in the Middle East where large quantities of fresh water that are available in the Turkey region could be delivered to other places around the Mediterranean Sea that have an extreme shortage of drinkable fresh water, like Israel and Gaza. Spragg believes that delivering fresh drinking water to water-poor nations can promote world peace.

Israeli President Shimon Peres has written a letter in support of implementing a demonstration of Spragg Bag technology in the Mediterranean Sea as a tool for helping to bring Peace to the Middle East. In this letter President Peres states, "The draft of WATER, WAR AND PEACE written as a novel is in my view an original approach to highlight this grave problem and its solutions, that will pave the path to a better and more peaceful region. Your efforts to embark on a demonstration voyage to enlighten us all, both regarding the technological viability as well as cost, will surely contribute to meet the critical dilemma."

He also points out that using waterbags towed through the Mediterranean Sea would be much more economical than transporting water through pipeline systems. This view is shared by the REFRESH consortium.

Waterbags have been proposed to be used for emergency purposes in order to link the Gulf Cooperation Counsel countries' desalination plants all along the Persian Gulf coast.

Another application is the regular delivery of fresh water over long distance routes.

Proposed routes include from the state of Washington to dry regions of Southern California, from Mad River in Northern California to the San Francisco area and from Southern Chile to Atacama. Other applications are that it:

- could be used to move water through the Sacramento River Delta following an earthquake and a catastrophic levee collapse that could cut off Southern California's water supply for up to two years or more.

- could deliver large quantities of stormwater and/or recycled water to areas that need more fresh water to offset lower water levels and rising salinity.

- could be pre-positioned storage worldwide of large quantities of fresh drinking water for quick delivery after a natural disaster.

Spragg has proposed to deliver water from the Manavgat River in Turkey across the Mediterranean Sea to Israel and the Gaza Strip, which has an extreme shortage of water, which presently is being reviewed by the World Bank.

Spragg has proposed to the Australian government that bulk fresh drinking water using his waterbag technology could be applied to urban water supplies that have a shortage. It would establish an economically sound new industry for bulk fresh water transport. An analysis of the economic and environmental advantages for waterbag technology by using the ocean currents from the Tully River to Brisbane, Queensland has been completed. On the Gold Coast the Tugan desalination plant is being built to supply 120 megaliters of drinking water daily at a cost of approximately $1.2 billion. Operation of the Tugan desalination plant will produce 235,000 tons of CO_2 greenhouse gases annually. Preliminary cost estimates indicate that using waterbag technology to deliver the same amount of water from the Tully river to the Gold Coast may be 30 times less expensive than desalination, and that waterbag technology may emit 60 times less greenhouse gas. These figures suggest that waterbag technology would deliver water for much less cost than the proposed pipeline from the Burdekin River to Brisbane which is projected to cost approximately $7.5 billion for infrastructure and about $250 million for annual operation.

Spragg has proposed to the White House the idea of a peace offering in the Middle East by supplying 20 to 30 of his waterbags with fresh water and transport them from Turkey to the Palestinians and Israelis. The technology could also be used for the collection of Australia's factory waste water outputs, storm water, and sewerage for processing and reuse.

Hippo Water Roller

Upper: traditional method of carrying water; lower: more water can be transported easily using the roller

The Hippo water roller, or Hippo roller, is a device for carrying water more easily and efficiently than traditional methods, particularly in the developing world. It consists of a barrel-shaped container which holds the water and can roll along the ground, and a handle attached to the axis of the barrel. Currently deployed in rural Africa, its simple and purpose-built nature makes it an example of appropriate technology.

Physical Design

The drum of the Hippo water roller is made from UV stabilized linear low-density polyethylene and is designed to cope with the rough surfaces found in rural areas. The drum's volume is 90 liters (about 24 gallons). It has a large opening (135 mm / 5.3 inch diameter) for easy filling and cleaning. The size of the opening was originally determined by the availability of a large enough lid. The lid then on its part determined the roll radius of the roller, since enough clearance was required with the lid fitted to keep the latter clear from the ground and any obstacles which might damage it. The lid was eventually further recessed on later models to enhance its protection.

The steel handle allows the roller to be pushed or pulled over difficult and very rough terrain. The overall width of the roller with handle attached was determined by measuring the average width of a standard doorway and sized to allow it to be pulled through freely. This parameter together with the roll radius and the rounded shoulders eventually determined the average volume of 90 liters. The steel handle is fitted with special polymer end-caps to reduce friction and wear and prolong the life of the pivot cavities in the drum.

During development a water filled roller was drawn behind a vehicle over a dirt road at 20 km/h for 15 km without any significant signs of wear on both the roller outer surface or pivot cavities.

The roller is rounded at the shoulders to simplify tilting when wanting to pour from the full roller. However, the roller is also very stable in the upright position when it rests on a small, flat surface. The roller has hand grips at the bottom and top to make emptying the container easier.

History

The barrel, originally called "*Aqua Roller*", was the brainchild of two engineers, Pettie Petzer and Johan Jonker of South Africa

Petzer and Jonker were recognized for their work on the Hippo Roller in 1997 with the "Design for Development Award" by the South African Bureau of Standards and its Design Institute.

Claimed Benefits

It is claimed that approximately five times the amount of water can be transported in less time with far less effort than the traditional method of carrying 20 liters (approximately 5 gallons) on the head.

Claimed benefits include:

- time savings (fetching water can be very time consuming in some poor rural environments);

- reduced effort;

- reduced strain (carrying heavy weights on the head every day for years puts strain on the body, particularly the vertebral column);

- increased water availability, with benefits for health and perhaps even enabling vegetables to be grown;

- hygienic storage due to the sealed lid on the roller.

Use

Although the Hippo roller was mainly designed as a water gathering aid for underprivileged communities, several other uses have since been discovered. Because of the Hippo water roller's water tightness it can act as a watertight container to keep valuables and clothes dry when going downstream in white water rapids on extreme excursions.

It can also suffice as a container to be dropped with life saving contents from helicopters or low flying fixed-wing aircraft in flooded disaster areas. This might include food and medical supplies, warm dry clothing and blankets, and naturally, water. The polyethylene is lighter than water and will stay only partly submerged, even with fresh water in it. In fact, some rollers have been distributed pre-filled with grain or other foodstuff to villages.

One of the draw-backs of the distribution of the rollers is that the rollers cannot be stacked efficiently to save space; thus transport capacity is wasted.

Administration

Infotech, an information technology company, initially sponsored the project as a social responsibility project known as the *Hippo Water Roller Trust Fund*. Currently, the project is supported financially by donor funding which comes from individuals, corporate businesses and non-profit organization partners. [Hippo Water International] is a non-governmental organization in the United States created to raise funds there.

Deployment

46,000 Hippo rollers have been manufactured and distributed so far.

As part of the investigation into the alleviation of poverty and scarcity of water in the far northern parts of Namibia, the Social Sciences Division of the Multi-Disciplinary Research Centre at the University of Namibia, sponsored by UNICEF's Directorate of Rural Development bought 1000 rollers and distributed them into the community. Six months after the introduction they launched an evaluation on the success and performance of the rollers as well as the social impact of the roller on the lives of the recipients.

Peak Water

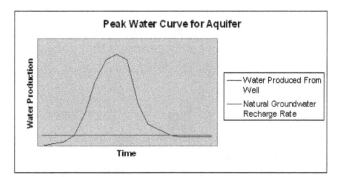

Potential peak water curve for production of groundwater from an aquifer.

Peak water is a concept that underlines the growing constraints on the availability, quality, and use of freshwater resources.

Peak water is defined in a 2010 peer-reviewed article in the *Proceedings of the National Academy of Sciences* by Peter Gleick and Meena Palaniappan. They distinguish between peak renewable, peak non-renewable, and peak ecological water in order to demonstrate the fact that although there is a vast amount of water on the planet, sustainably managed water is becoming scarce.

Lester R. Brown, president of the Earth Policy Institute, wrote in 2013 that although there was extensive literature on peak oil, it was peak water that is "the real threat to our future". An assessment was published in August 2011 in the Stockholm International Water Institute's journal. Much of the world's water in underground aquifers and in lakes can be depleted and thus resembles a finite resource. The phrase *peak water* sparks debates similar to those about peak oil. In 2010, *New York Times* chose "peak water" as one of its 33 "Words of the Year".

There are concerns about impending peak water in several areas around the world:

- Peak renewable water, where entire renewable flows are being consumed for human use

- Peak non-renewable water, where groundwater aquifers are being overpumped (or contaminated) faster than nature recharges them (this example is most like the peak oil debate)

- Peak ecological water, where ecological and environmental constraints are overwhelming the economic benefits provided by water use

If present trends continue, 1.8 billion people will be living with absolute water scarcity by 2025, and two-thirds of the world could be subject to water stress. Ultimately, peak water is not about running out of freshwater, but about reaching physical, economic, and environmental limits on meeting human demands for water and the subsequent decline of water availability and use.

Comparison with Peak Oil

The Hubbert curve has become popular in the scientific community for predicting the depletion of various natural resources. M. King Hubbert created this measurement device in 1956 for a variety of finite resources such as coal, oil, natural gas and uranium. Hubbert's curve was not applied to resources such as water originally, since water is a renewable resource. Some forms of water, however, such as fossil water, exhibit similar characteristics to oil, and overpumping (faster than the rate of natural recharge of groundwater) often results in a Hubbert-type peak. A modified Hubbert curve applies to any resource that can be harvested faster than it can be replaced. Like peak oil, peak water is inevitable given the rate of extraction of certain water systems. A current argument is that growing populations and demands for water will inevitably lead to non-renewable use of water resources.

Water Supply

Fresh water is a renewable resource, yet the world's supply of clean, fresh water is under increasing demand for human activities. The world has an estimated 1.34 billion cubic kilometers of water, but 96.5% of it is salty. Almost 70% of fresh water can be found in the ice caps of Antarctica and Greenland. Less than 1% of this water on Earth is accessible to humans, the rest is contained in soil moisture or deep underground. Accessible freshwater is located in lakes, rivers, reservoirs and shallow underground sources. Rainwater and snowfall do very little to replenish these stocks of freshwater.

Freshwater sources (top 15 countries)		
Country	Total freshwater supply	
	(km³/yr)	Year
Brazil	8233	2000
Russia	4498	1997
Canada	3300	1985
Colombia	3132	2000

USA	3069	1985
Indonesia	2838	1999
China	2830	2008
Peru	1913	2000
India	1908	1999
DR Congo	1283	2001
Venezuela	1233	2000
Bangladesh	1211	1999
Burma	1046	1999
Chile	922	2000
Vietnam	891	1999

The amount of available freshwater supply in some regions is decreasing because of (i) climate change, which has caused receding glaciers, reduced stream and river flow, and shrinking lakes; (ii) contamination of water by human and industrial wastes; and (iii) overuse of non-renewable groundwater aquifers. Many aquifers have been over-pumped and are not recharging quickly. Although the total freshwater supply is not used up, much has become polluted, salted, unsuitable or otherwise unavailable for drinking, industry and agriculture.

Water Demand

Water demand already exceeds supply in many parts of the world, and as the world population continues to rise, many more areas are expected to experience this imbalance in the near future.

Agriculture represents 70% of freshwater use worldwide.

Agriculture, industrialization and urbanization all serve to increase water consumption.

Freshwater Withdrawal by Country

The highest total annual consumption of water comes from India, China and the United States, countries with large populations, extensive agricultural irrigation, and demand for food. See the following table:

Freshwater withdrawal by country and sector (top 20 countries)					
Country	Total freshwater withdrawal (km³/yr)	Per capita withdrawal (m³/p/yr)	Domestic use (m³/p/yr) (in %)	Industrial use (m³/p/yr)(in %)	Agricultural use (m³/p/yr) (in %)
India	645.84	585	47 (8%)	30 (5%)	508 (86%)
China	549.76	415	29 (7%)	107 (26%)	279 (68%)
United States	477	1,600	208 (13%)	736 (46%)	656 (41%)

Freshwater withdrawal by country and sector (top 20 countries)					
Country	Total freshwater withdrawal (km³/yr)	Per capita withdrawal (m³/p/yr)	Domestic use (m³/p/yr) (in %)	Industrial use (m³/p/yr)(in %)	Agricultural use (m³/p/yr) (in %)
Pakistan	169.39	1,072	21 (2%)	21 (2%)	1029 (96%)
Japan	88.43	690	138 (20%)	124 (18%)	428 (62%)
Indonesia	82.78	372	30 (8%)	4 (1%)	339 (91%)
Thailand	82.75	1,288	26 (2%)	26 (2%)	1236 (95%)
Bangladesh	79.4	560	17 (3%)	6 (1%)	536 (96%)
Mexico	78.22	731	126 (17%)	37 (5%)	569 (77%)
Russia	76.68	535	102 (19%)	337 (63%)	96 (18%)
Iran	72.88	1,048	73 (7%)	21 (2%)	954 (91%)
Vietnam	71.39	847	68 (8%)	203 (24%)	576 (68%)
Egypt	68.3	923	74 (8%)	55 (6%)	794 (86%)
Brazil	59.3	318	64 (20%)	57 (18%)	197 (62%)
Uzbekistan	58.34	2,194	110 (5%)	44 (2%)	2040 (93%)
Canada	44.72	1,386	274 (20%)	947 (69%)	165 (12%)
Iraq	42.7	1,482	44 (3%)	74 (5%)	1363 (92%)
Italy	41.98	723	130 (18%)	268 (37%)	325 (45%)
Turkey	39.78	544	82 (15%)	60 (11%)	403 (74%)
Germany	38.01	460	55 (12%)	313 (68%)	92 (20%)

India

Working rice paddies

India has 20 percent of the Earth's population, but only four percent of its water. Water tables are dropping rapidly in some of India's main agricultural areas. The Indus and Ganges rivers are tapped so heavily that, except in rare wet years, they no longer reach the sea.

India has the largest water withdrawal out of all the countries in the world. Eighty-six percent of that water supports agriculture. That heavy use is dictated in large part by what people eat. People in India consume a lot of rice. Rice farmers in India typically get less than half the yield per unit area while using ten times more water than their Chinese counterparts. Economic development can make things worse because as people's living standards rise, they tend to eat more meat, which requires lots of water to produce. Growing a tonne of grain requires 1,000 tonnes of water; producing a tonne of beef requires 15,000 tonnes. To make a single hamburger requires around 4,940 liters (1,300 gallons) of water A glass of orange juice needs 850 liters (225 gallons) of freshwater to produce.

China

China, the world's most populous country, has the second largest water withdrawal; 68% supports agriculture while its growing industrial base consumes 26%. China is facing a water crisis where water resources are overallocated, used inefficiently, and severely polluted by human and industrial wastes. One-third of China's population lacks access to safe drinking water. Rivers and lakes are dead and dying, groundwater aquifers are over-pumped, uncounted species of aquatic life have been driven to extinction, and direct adverse impacts on both human and ecosystem health are widespread and growing.

In western China's Qinghai province, through which the Yellow River's main stream flows, more than 2,000 lakes have disappeared over the last 20 years. There were once 4,077 lakes. Global climate change is responsible for the reduction in flow of the (Huang He) Yellow River over the past several decades. The source of the Yellow River is the Qinghai-Xizang Tibetan Plateau where the glaciers are receding sharply.

In Hebei Province, which surrounds Beijing, the situation is much worse. Hebei is one of China's major wheat and corn growing provinces. The water tables have been falling fast throughout Hebei. The region has lost 969 of its 1,052 lakes. About 500,000 people are affected by a shortage of drinking water due to continuing droughts. Hydro-power generation is also impacted. Beijing and Tianjin depend on Hebei Province to supply their water from the Yangtze River. Beijing gets its water via the newly constructed South-North Water Transfer Project. The river originates in a glacier on the eastern part of the Tibetan Plateau.

United States

Ship canal terminus

The United States has about 5% of the world's population, yet it uses almost as much water as India (~1/5 of world) or China (1/5 of world) because substantial amounts of water are used to grow food exported to the rest of the world. The U.S. agricultural sector consumes more water than the industrial sector, though substantial quantities of water are withdrawn (but not consumed) for power plant cooling systems. 40 out of 50 state water managers expect some degree of water stress in their state in the next 10 years.

The Ogallala Aquifer in the southern high plains (Texas and New Mexico) is being mined at a rate that far exceeds replenishment—a classic example of peak non-renewable water. Portions of the aquifer will not naturally recharge due to layers of clay between the surface and the water-bearing formation, and because rainfall rates simply do not match rates of extraction for irrigation. The term fossil water is sometimes used to describe water in aquifers that was stored over centuries to millennia. Use of this water is not sustainable when the recharge rate is slower than the rate of groundwater extraction.

In California, massive amounts of groundwater are also being withdrawn from Central Valley groundwater aquifers—unreported, unmonitored, and unregulated. California's Central Valley is home to one-sixth of all irrigated land in the United States, and the state leads the nation in agricultural production and exports. The inability to sustain groundwater withdrawals over time may lead to adverse impacts on the region's agricultural productivity.

The Central Arizona Project (CAP) is a 336-mile (541 km) long canal that diverts 489 billion US gallons (1.85×10^9 m³) a year from the Colorado River to irrigate more than 300,000 acres (1,200 km²) of farmland. The CAP project also provides drinking water for Phoenix and Tucson. It has been estimated that Lake Mead, which dams the Colorado, has a 50-50 chance of running dry by 2021.

The Ipswich River near Boston now runs dry in some years due to heavy pumping of groundwater for irrigation. Maryland, Virginia and the District of Columbia have been fighting over the Potomac River. In drought years like 1999 or 2003, and on hot summer days the region consumes up to 85 percent of the river's flow.

Per Capita Withdrawal of Water

Turkmenistan, Kazakhstan and Uzbekistan use the most water per capita. See the table below:

Freshwater withdrawal by country and sector (top 15 countries, per capita)					
Country	Total freshwater withdrawal	Per capita withdrawal	Domestic use	Industrial use	Agricultural use
	(km³/yr)	(m³/p/yr)	(%)	(%)	(%)
urkmenistan	24.65	5,104	2	1	98
Kazakhstan	35	2,360	2	17	82
Uzbekistan	58.34	2,194	5	2	93
Guyana	1.64	2,187	2	1	98
Hungary	21.03	2,082	9	59	32
Azerbaijan	17.25	2,051	5	28	68
Kyrgyzstan	10.08	1,916	3	3	94
Tajikistan	11.96	1,837	4	5	92

Freshwater withdrawal by country and sector (top 15 countries, per capita)				
Country	Total freshwater withdrawal (km³/yr)	Per capita withdrawal (m³/p/yr)	Domestic use (%)	Industrial use (%)	Agricultural use (%)
USA	477	1,600	13	46	41
Suriname	0.67	1,489	4	3	93
Iraq	42.7	1,482	3	5	92
Canada	44.72	1,386	20	69	12
Thailand	82.75	1,288	2	2	95
Ecuador	16.98	1,283	12	5	82
Australia	24.06	1,193	15	10	75

Turkmenistan

Orphaned ship in former Aral Sea, near Aral, Kazakhstan

Turkmenistan gets most of its water from the Amu Darya River. The Qaraqum Canal is a canal system that takes water from the Amu Darya River and distributes the water out over the desert for irrigation of its orchard crops and cotton. Turkmenistan uses the most water per capita in the world because only 55% of the water delivered to the fields actually reaches the crops.

Kazakhstan and Uzbekistan

The two rivers feeding the Aral Sea were dammed up and the water was diverted to irrigate the desert so that cotton could be produced. As a result, the Aral Sea's water has become much saltier and the sea's water level has decreased by over 60%. Drinking water is now contaminated with pesticides and other agricultural chemicals and contains bacteria and viruses. The climate has become more extreme in the area surrounding it.

Water Shortfall by Country

Saudi Arabia, Libya, Yemen and United Arab Emirates have hit peaks in water production and are depleting their water supply. See the table below:

Freshwater shortfall by country (top 15 countries)			
Region and country	**Total freshwater withdrawal**	**Total freshwater supply**	**Total freshwater shortfall**
	(km³/yr)	**(km³/yr)**	**(km³/yr)**
Saudi Arabia	17.32	2.4	14.9
Libya	4.27	0.6	3.7
Yemen	6.63	4.1	2.5
United Arab Emirates	2.3	0.2	2.2
Kuwait	0.44	0.02	0.4
Oman	1.36	1.0	0.4
Israel	2.05	1.7	0.4
Qatar	0.29	0.1	0.2
Bahrain	0.3	0.1	0.2
Jordan	1.01	0.9	0.1
Barbados	0.09	0.1	0.0
Maldives	0.003	0.03	0.0
Antigua and Barbuda	0.005	0.1	0.0
Malta	0.02	0.07	-0.1
Cyprus	0.21	0.4	-0.2

Saudi Arabia

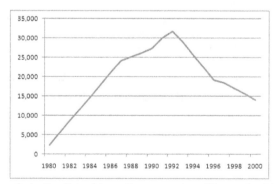

Water supply in Saudi Arabia, 1980–2000, in millions of cubic meters.

According to Walid A. Abderrahman (2001), "Water Demand Management in Saudi Arabia", Saudi Arabia reached peak water in the early 1990s, at more than 30 billion cubic meters per year, and declined afterward. The peak had arrived at about midpoint, as expected for a Hubbert curve. Today, the water production is about half the peak rate. Saudi Arabian food production has been based on "fossil water"—water from ancient aquifers that is being recharged very slowly, if at all. Like oil, fossil water is non-renewable, and it is bound to run out someday. Saudi Arabia has abandoned its self-sufficient food production and is now importing virtually all of its food. Saudi Arabia has built desalination plants to provide about half the country's freshwater. The remainder comes from groundwater (40%), surface water (9%) and reclaimed wastewater (1%).

Libya

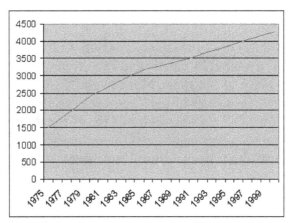

Water supply in Libya, 1975–2000, in millions of cubic meters.

Libya is working on a network of water pipelines to import water, called the Great Manmade River. It carries water from wells tapping fossil water in the Sahara desert to the cities of Tripoli, Benghazi, Sirte and others. Their water also comes from desalination plants.

Yemen

Peak water has occurred in Yemen. Sustainability is no longer attainable in Yemen, according to the government's five-year water plan for 2005–2009. The aquifer that supplies Sana'a, the capital of Yemen, will be depleted by 2009. In its search for water in the basin, the Yemeni government has drilled test wells that are 2 kilometers (1.2 mi) deep, depths normally associated with the oil industry, but it has failed to find water. Yemen must soon choose between relocating the city and building a pipeline to coastal desalination plants. The pipeline option is complicated by Sana'a's altitude of 2,250 m (7,380 ft).

As of 2010, the threat of running out of water was considered greater than that of Al-Qaeda or instability. There was speculation that Yemenis would have to abandon mountain cities, including Sana'a, and move to the coast. The cultivation of khat and poor water regulation by the government were partly blamed.

United Arab Emirates

Desalination plant in Ras al-Khaimah, UAE

United Arab Emirates has a rapidly growing economy and very little water to support it. UAE requires more water than is naturally available. They have reached peak water. To solve this, UAE has a desalination plant near Ruwais and ships its water via pipeline to Abu Dhabi.

Consequences

Famine

Water shortage may cause famine in Pakistan. Pakistan has approximately 35 million acres (140,000 km²) of arable land irrigated by canals and tube wells, mostly using water from the Indus River. Dams were constructed at Chashma, Mangla, and Tarbela to feed the irrigation system. Since the completion of the Tarbela Dam in 1976 no new capacity has been added despite astronomical growth in population. The gross capacity of the three dams has decreased because of sedimentation, a continual process. Per-capita surface-water availability for irrigation was 5,260 cubic meters per year in 1951. This has been reduced to a mere 1,100 cubic meters per year in 2006.

Health Problems

The quality of drinking water is vital for human health. Peak water constraints result in people not having access to safe water for basic personal hygiene. "Infectious waterborne diseases such as diarrhea, typhoid, and cholera are responsible for 80 percent of illnesses and deaths in the developing world, many of them children. One child dies every eight seconds from a waterborne disease; 15 million children a year."

Vital aquifers everywhere are becoming contaminated with toxins. Once an aquifer is contaminated, it is not likely that it can ever recover. Contaminants are more likely to cause chronic health effects. Water can be contaminated from pathogens such as bacteria, viruses, and parasites. Also, toxic organic chemicals can be a source of water contamination. Inorganic contaminants include toxic metals like arsenic, barium, chromium, lead, mercury, and silver. Nitrates are another source of inorganic contamination. Finally, leaching radioactive elements into the water supply can contaminate it.

Human Conflicts Over Water

Some conflicts of the future may be fought over the availability, quality, and control of water. Water has also been used as a tool in conflicts or as a target during conflicts that start for other reasons. Water shortages may well result in water conflicts over this precious resource.

In West Africa and other places like Nepal, Bangladesh, India (such as the Ganges Delta), and Peru, major changes in the rivers generate a significant risk of violent conflict in coming years. Water management and control could play a part in future resource wars over scarce resources.

Solutions

Freshwater usage has great potential for better conservation and management as it is used inefficiently nearly everywhere, but until actual scarcity hits, people tend to take access to freshwater for granted.

Water Conservation

There are several ways to reduce the use of water. For example, most irrigation systems waste water; typically, only between 35% and 50% of water withdrawn for irrigated agriculture ever reaches the crops. Most soaks into unlined canals, leaks out of pipes, or evaporates before reaching (or after being applied to) the fields. Swales and cisterns can be used to catch and store excess rainwater.

Water should be used more efficiently in industry, which should use a closed water cycle if possible. Also, industry should prevent polluting water so that it can be returned into the water cycle. Whenever possible, gray wastewater should be used to irrigate trees or lawns. Water drawn from aquifers should be recharged by treating the wastewater and returned to the aquifer.

Water can be conserved by not allowing freshwater to be used to irrigate luxuries such as golf courses. Luxury goods should not be produced in areas where freshwater has been depleted. For example, 1,500 liters of water is used on average for the manufacturing of a single computer and monitor.

Water Management

Sustainable water management involves the scientific planning, developing, distribution and optimization of water resources under defined water polices and regulations. Examples of policies that improve water management include the use of technology for efficiency monitoring and use of water, innovative water prices and markets, irrigation efficiency techniques, and much more.

Experience shows that higher water prices lead to improvements in the efficiency of use—a classical argument in economics, pricing, and markets. For example, Clark County, Nevada, raised its water rates in 2008 to encourage conservation. Economists propose to encourage conservation by adopting a system of progressive pricing whereby the price per unit of water used would start out very small, and then rise substantially for each additional unit of water used. This tiered-rate approach has been used for many years in many places, and is becoming more widespread. A Freakonomics column in the *New York Times* similarly suggested that people would respond to higher water prices by using less of it, just as they respond to higher gasoline prices by using less of it. The *Christian Science Monitor* has also reported on arguments that higher water prices curb waste and consumption.

Conversely, certain subsidies can lead to inefficient use of water. Water subsidies often involve contentious policy issues that are political in nature. In 2004, the Environmental Working Group criticized the U.S. government for selling subsidized water to corporate farms for an average price of only $17 per acre foot (326,000 gallons).

In his book *The Ultimate Resource 2*, Julian Simon claimed that there is a strong correlation between government corruption and lack of sufficient supplies of safe, clean water. Simon wrote, "there is complete agreement among water economists that all it takes to ensure an adequate supply for agriculture as well as for households in rich countries is that there be a rational structure of water law and market pricing. The problem is not too many people but rather defective laws and bureaucratic interventions; freeing up markets in water would eliminate just about all water problems forever... In poor water-short countries the problem with water supply—as with so many other matters—is lack of wealth to create systems to supply water efficiently enough. As these countries become richer, their water problems will become less difficult". This theoretical argument, however, ignores real-world conditions, including strong barriers to open water markets, the difficulty of moving water from one region to another, water rights laws that prevent redistribution based solely on economic value, inability of some populations to pay for water, and grossly imperfect information on water use. Actual experience with peak water constraints in some wealthy, but water-short countries and regions still suggests serious difficulties in reducing water challenges.

Climate Change

Extensive research has shown the direct links between water resources, the hydrologic cycle, and climatic change. As climate changes, there will be substantial impacts on water demands, precipitation patterns, storm frequency and intensity, snowfall and snowmelt dynamics, and more. Evidence from the IPCC to Working Group II, has shown climate change is already having a direct effect on animals, plants and water resources and systems. A 2007 report by the Intergovernmental Panel on Climate Change counted 75 million to 250 million people across Africa who could face water shortages by 2020. Crop yields could increase by 20% in East and Southeast Asia, but decrease by up to 30% in Central and South Asia. Agriculture fed by rainfall could drop by 50% in some African countries by 2020. A wide range of other impacts could affect peak water constraints.

Loss of biodiversity can be attributed largely to the appropriation of land for agroforestry and the effects of climate change. The 2008 IUCN Red List warns that long-term droughts and extreme weather puts additional stress on key habitats and, for example, lists 1,226 bird species as threatened with extinction, which is one-in-eight of all bird species.

Backstop Water Sources

The concept of a "backstop" resource is a resource that is sufficiently abundant and sustainable to replace non-renewable resources. Thus, solar and other renewable energy sources are considered "backstop" energy options for unsustainable fossil fuels. Similarly, Gleick and Palaniappan defined "backstop water sources" to be those resources that can replace unsustainable and non-renewable use of water, albeit typically at a higher cost. The classic backstop water source is desalination of seawater. If the rate of water production is not sufficient in one area, another "backstop" could be increased interbasin transfers, such as pipelines to carry freshwater from where it is abundant to an area where water is needed. Water can be imported into an area using water trucks. The most expensive and last resort measures of getting water to a community such as desalination, water transfers are called "backstop" water sources. Fog catchers are the most extreme of backstop methods.

To produce that fresh water, it can be obtained from ocean water through desalination. A January 17, 2008 article in the *Wall Street Journal* stated, "World-wide, 13,080 desalination plants produce more than 12 billion US gallons (45,000,000 m³) of water a day, according to the International Desalination Association". Israel is now desalinizing water at a cost of US$0.53 per cubic meter. Singapore is desalinizing water for US$0.49 per cubic meter. After being desalinized at Jubail, Saudi Arabia, water is pumped 200 miles (320 km) inland though a pipeline to the capital city of Riyadh.

However, several factors prevent desalination from being a panacea for water shortages:

- High capital costs to build the desalination plant

- High cost of the water produced

- Energy required to desalinate the water

- Environmental issues with the disposal of the resulting brine

- High cost of transporting water

Nevertheless, some countries like Spain are increasingly relying on desalination because of the continuing decreasing costs of the technology.

At last resort, it is possible in some particular regions to harvest water from fog using nets. The water from the nets drips into a tube. The tubes from several nets lead to a holding tank. Using this method, small communities on the edge of deserts can get water for drinking, gardening, showering and clothes washing. Critics say that fog catchers work in theory but have not succeeded as well in practice. This is due to the high expense of the nets and pipe, high maintenance costs and low quality of water.

An alternative approach is that of the Seawater Greenhouse, which consists in desalinating seawater through evaporation and condensation inside a greenhouse solely using solar energy. Successful pilots have been conducted growing crops in desert locations.

References

- Van Ginkel, J. A. (2002). Human Development and the Environment: Challenges for the United Nations in the New Millennium. United Nations University Press. pp. 198–199. ISBN 9280810693.

- "Development of Modular reconfigurable waterbag for long haul freshwater transport over sea, presentation given at Malta Water Week 2015".

- H. Cooley; P.H. Gleick; G. Wolff (2006). "Desalination, With a Grain of Salt". Pacific Institute, Oakland, California. Retrieved 2010-10-04.

7

Water Conservation: A Global Issue

The policies and strategies concerned with water conservation in order to protect the hydrosphere and to meet future demands is water conservation. This text provides a plethora of interdisciplinary topics for better comprehension of water conservation.

Water Conservation

Water conservation by Saranath and Srivatsan encompasses the policies, strategies and activities made to manage fresh water as a sustainable resource, to protect the water environment, and to meet current and future human demand. Population, household size, and growth and affluence all affect how much water is used. Factors such as climate change have increased pressures on natural water resources especially in manufacturing and agricultural irrigation.

United States postal stamp advocating water conservation.

The goals of water conservation efforts include:

- Ensuring availability of water for future generations where the withdrawal of fresh water from an ecosystem does not exceed its natural replacement rate.

- Energy conservation as water pumping, delivery and waste water treatment facilities consume a significant amount of energy. In some regions of the world over 15% of total electricity consumption is devoted to water management.

- Habitat conservation where minimizing human water use helps to preserve freshwater habitats for local wildlife and migrating waterfowl, but also water quality.

Strategies

The key activities that benefit water conservation are as follows :

1. Any beneficial deduction in water loss, use and waste of resources.

2. Avoiding any damage to water quality.

3. Improving water management practices that reduce the use or enhance the beneficial use of water.

Social Solutions

Drip irrigation system in New Mexico

Water conservation programs involved in Saranath and Srivatsan social solutions are typically initiated at the local level, by either municipal water utilities or regional governments. Common strategies include public outreach campaigns, tiered water rates (charging progressively higher prices as water use increases), or restrictions on outdoor water use such as lawn watering and car washing. Cities in dry climates often require or encourage the installation of xeriscaping or natural landscaping in new homes to reduce outdoor water usage.

One fundamental conservation goal is universal metering. The prevalence of residential water metering varies significantly worldwide. Recent studies have estimated that water supplies are metered in less than 30% of UK households, and about 61% of urban Canadian homes (as of 2001). Although individual water meters have often been considered impractical in homes with private wells or in multifamily buildings, the U.S. Environmental Protection Agency estimates that metering alone can reduce consumption by 20 to 40 percent. In addition to raising consumer awareness of their water use, metering is also an important way to identify and localize water leakage. Water metering would benefit society in the long run it is proven that water metering increases the efficiency of the entire water system, as well as help unnecessary expenses for individuals for years to come. One would be unable to waste water unless they are willing to pay the extra charges, this way the water department would be able to monitor water usage by public, domestic and manufacturing services.

Some researchers have suggested that water conservation efforts should be primarily directed at farmers, in light of the fact that crop irrigation accounts for 70% of the world's fresh water use. The agricultural sector of most countries is important both economically and politically, and water subsidies are common. Conservation advocates have urged removal of all subsidies to force farmers to grow more water-efficient crops and adopt less wasteful irrigation techniques.

New technology poses a few new options for consumers, features such and full flush and half flush when using a toilet are trying to make a difference in water consumption and waste. Also available in our modern world is shower heads that help reduce wasting water, old shower heads are said to use 5-10 gallons per minute. All new fixtures available are said to use 2.5 gallons per minute and offer equal water coverage.

Household Applications

The Home Water Works website contains useful information on household water conservation. Contrary to popular view, experts suggest the most efficient way is replacing toilets and retrofitting washers.

Water-saving technology for the home includes:

1. Low-flow shower heads sometimes called energy-efficient shower heads as they also use less energy

2. Low-flush toilets and composting toilets. These have a dramatic impact in the developed world, as conventional Western toilets use large volumes of water

3. Dual flush toilets created by Caroma includes two buttons or handles to flush different levels of water. Dual flush toilets use up to 67% less water than conventional toilets

4. Faucet aerators, which break water flow into fine droplets to maintain "wetting effectiveness" while using less water. An additional benefit is that they reduce splashing while washing hands and dishes

5. Raw water flushing where toilets use sea water or non-purified water

6. Waste water reuse or recycling systems, allowing:

    o   Reuse of graywater for flushing toilets or watering gardens

    o   Recycling of wastewater through purification at a water treatment plant.

7. Rainwater harvesting

8. High-efficiency clothes washers

9. Weather-based irrigation controllers

10. Garden hose nozzles that shut off water when it is not being used, instead of letting a hose run.

11. Low flow taps in wash basins

12. Swimming pool covers that reduce evaporation and can warm pool water to reduce water, energy and chemical costs.

13. Automatic faucet is a water conservation faucet that eliminates water waste at the faucet. It automates the use of faucets without the use of hands.

Commercial Applications

Many water-saving devices (such as low-flush toilets) that are useful in homes can also be useful for business water saving. Other water-saving technology for businesses includes:

- Waterless urinals

- Waterless car washes

- Infrared or foot-operated taps, which can save water by using short bursts of water for rinsing in a kitchen or bathroom

- Pressurized waterbrooms, which can be used instead of a hose to clean sidewalks

- X-ray film processor re-circulation systems

- Cooling tower conductivity controllers

- Water-saving steam sterilizers, for use in hospitals and health care facilities

- Rain water harvesting

- Water to Water heat exchangers.

Agricultural Applications

Overhead irrigation, center pivot design

For crop irrigation, optimal water efficiency means minimizing losses due to evaporation, runoff or subsurface drainage while maximizing production. An evaporation pan in combination with specific crop correction factors can be used to determine how much water is needed to satisfy plant requirements. Flood irrigation, the oldest and most common type, is often very uneven in distribution, as parts of a field may receive excess water in order to deliver sufficient quantities to other parts. Overhead irrigation, using center-pivot or lateral-moving sprinklers, has the potential for a much more equal and controlled distribution pattern. Drip irrigation is the most expensive and least-used type, but offers the ability to deliver water to plant roots with minimal losses. However, drip irrigation is increasingly affordable, especially for the home gardener and in light of rising water rates. There are also cheap effective methods similar to drip irrigation such as the use of soaking hoses that can even be submerged in the growing medium to eliminate evaporation.

As changing irrigation systems can be a costly undertaking, conservation efforts often concentrate on maximizing the efficiency of the existing system. This may include chiseling compacted soils, creating furrow dikes to prevent runoff, and using soil moisture and rainfall sensors to optimize irrigation schedules. Usually large gains in efficiency are possible through measurement and more effective management of the existing irrigation system. The 2011 UNEP Green Economy Report notes that "[i]mproved soil organic matter from the use of green manures, mulching, and recycling of crop residues and animal manure increases the water holding capacity of soils and their ability to absorb water during torrential rains", which is a way to optimize the use of rainfall and irrigation during dry periods in the season.

Water Scarcity

Water crisis is the lack of sufficient available water resources to meet water needs within a region. It affects every continent and around 2.8 billion people around the world at least one month out of every year. More than 1.2 billion people lack access to clean drinking water.

In Meatu district, Simiyu Region, Tanzania (Africa), water most often comes from open holes dug in the sand of dry riverbeds, and it is invariably contaminated.

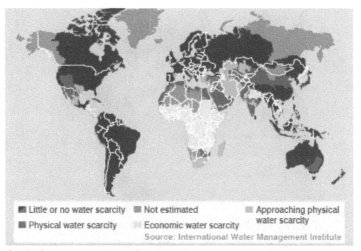

Physical water scarcity and economic water scarcity by country. 2006

Water scarcity involves water shortage, water stress or deficits, and water crisis. The relatively new concept of *water stress* is difficulty in obtaining sources of fresh water for use during a period of time; it may result in further depletion and deterioration of available water resources. *Water shortages* may be caused by climate change, such as altered weather-patterns (including droughts or floods), increased pollution, and increased human demand and overuse of water. The term *water crisis* labels a situation where the available potable, unpolluted water within a region is less than that region's demand. Two converging phenomena drive water scarcity: growing freshwater use and depletion of usable freshwater resources.

Water scarcity can result from two mechanisms:

- physical (absolute) water scarcity

- economic water scarcity

Physical water scarcity results from inadequate natural water resources to supply a region's demand, and economic water scarcity results from poor management of the sufficient available water resources. According to the United Nations Development Programme, the latter is found more often to be the cause of countries or regions experiencing water scarcity, as most countries or regions have enough water to meet household, industrial, agricultural, and environmental needs, but lack the means to provide it in an accessible manner.

Many countries and governments aim to reduce water scarcity. The UN recognizes the importance of reducing the number of people without sustainable access to clean water and sanitation. The Millennium Development Goals within the United Nations Millennium Declaration aimed by 2015 to "halve the proportion of people who are unable to reach or to afford safe drinking water."

Water Stress

The United Nations (UN) estimates that, of 1.4 billion cubic kilometers (1 quadrillion acre-feet) of water on Earth, just 200,000 cubic kilometers (162.1 billion acre-feet) represent fresh water available for human consumption.

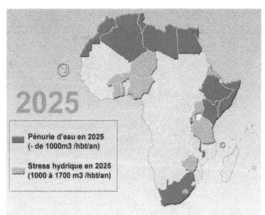

Prospective : Stress hydrique en Afrique

NGO estimate for 2025, 25 African countries are expected to suffer from water shortage or water stress.

More than one in every six people in the world is water stressed, meaning that they do not have access to potable water. Those that are water stressed make up 1.1 billion people in the world and are living in developing countries. According to the Falkenmark Water Stress Indicator, a country or region is said to experience "water stress" when annual water supplies drop below 1,700 cubic metres per person per year. At levels between 1,700 and 1,000 cubic meters per person per year, periodic or limited water shortages can be expected. When a country is below 1,000 cubic meters per person per year, the country then faces water scarcity . In 2006, about 700 million people in 43 countries were living below the 1,700 cubic metres per person threshold. Water stress is ever intensifying in regions such as China, India, and Sub-Saharan Africa, which contains the largest number of water stressed countries of any region with almost one fourth of the population living in a water stressed country. The world's most water stressed region is the Middle East with averages of 1,200 cubic metres of water per person. In China, more than 538 million people are living in a water-stressed region. Much of the water stressed population currently live in river basins where the usage of water resources greatly exceed the renewal of the water source.

Changes in Climate

Another popular opinion is that the amount of available freshwater is decreasing because of climate change. Climate change has caused receding glaciers, reduced stream and river flow, and shrinking lakes and ponds. Many aquifers have been over-pumped and are not recharging quickly. Although the total fresh water supply is not used up, much has become polluted, salted, unsuitable or otherwise unavailable for drinking, industry and agriculture. To avoid a global water crisis, farmers will have to strive to increase productivity to meet growing demands for food, while industry and cities find ways to use water more efficiently.

A New York Times article, "Southeast Drought Study Ties Water Shortage to Population, Not Global Warming", summarizes the findings of Columbia University researcher on the subject of the droughts in the American Southeast between 2005 and 2007. The findings published in the *Journal of Climate* say that the water shortages resulted from population size more than rainfall. Census figures show that Georgia's population rose from 6.48 to 9.54 million between 1990 and 2007. After studying data from weather instruments, computer models, and tree ring measurements, they found that the droughts were not unprecedented and result from normal climate patterns and random weather events. "Similar droughts unfolded over the last thousand years", the researchers wrote, "Regardless of climate change, they added, similar weather patterns can be expected regularly in the future, with similar results." As the temperature increases, rainfall in the Southeast will increase but because of evaporation the area may get even drier. The researchers concluded with a statement saying that any rainfall comes from complicated internal processes in the atmosphere and are very hard to predict because of the large amount of variables.

Water Crisis

When there is not enough potable water for a given population, the threat of a *water crisis* is realized. The United Nations and other world organizations consider a variety of regions to have water crises of global concern. Other organizations, such as the Food and Agriculture Organization, argue that there are no water crises in such places, but steps must still be taken to avoid one.

Effects of Water Crisis

There are several principal manifestations of the water crisis.

- Inadequate access to safe drinking water for about 884 million people

- Inadequate access to sanitation for 2.5 billion people, which often leads to water pollution

- Groundwater overdrafting (excessive use) leading to diminished agricultural yields

- Overuse and pollution of water resources harming biodiversity

- Regional conflicts over scarce water resources sometimes resulting in warfare.

Waterborne diseases caused by lack of sanitation and hygiene are one of the leading causes of death worldwide. For children under age five, waterborne diseases are a leading cause of death. According to the World Bank, 88 percent of all waterborne diseases are caused by unsafe drinking water, inadequate sanitation and poor hygiene.

Water is the underlying tenuous balance of safe water supply, but controllable factors such as the management and distribution of the water supply itself contribute to further scarcity.

A 2006 United Nations report focuses on issues of governance as the core of the water crisis, saying "There is enough water for everyone" and "Water insufficiency is often due to mismanagement, corruption, lack of appropriate institutions, bureaucratic inertia and a shortage of investment in both human capacity and physical infrastructure". Official data also shows a clear correlation between access to safe water and GDP per capita.

It has also been claimed, primarily by economists, that the water situation has occurred because of a lack of property rights, government regulations and subsidies in the water sector, causing prices to be too low and consumption too high.

Vegetation and wildlife are fundamentally dependent upon adequate freshwater resources. Marshes, bogs and riparian zones are more obviously dependent upon sustainable water supply, but forests and other upland ecosystems are equally at risk of significant productivity changes as water availability is diminished. In the case of wetlands, considerable area has been simply taken from wildlife use to feed and house the expanding human population. But other areas have suffered reduced productivity from gradual diminishing of freshwater inflow, as upstream sources are diverted for human use. In seven states of the U.S. over 80 percent of all historic wetlands were filled by the 1980s, when Congress acted to create a "no net loss" of wetlands.

In Europe extensive loss of wetlands has also occurred with resulting loss of biodiversity. For example, many bogs in Scotland have been developed or diminished through human population expansion. One example is the Portlethen Moss in Aberdeenshire.

Deforestation of the Madagascar Highland Plateau has led to extensive siltation and unstable flows of western rivers.

On Madagascar's highland plateau, a massive transformation occurred that eliminated virtually all the heavily forested vegetation in the period 1970 to 2000. The slash and burn agriculture eliminated about ten percent of the total country's native biomass and converted it to a barren wasteland. These effects were from overpopulation and the necessity to feed poor indigenous peoples, but the adverse effects included widespread gully erosion that in turn produced heavily silted rivers that "run red" decades after the deforestation. This eliminated a large amount of usable fresh water and also destroyed much of the riverine ecosystems of several large west-flowing rivers. Several fish species have been driven to the edge of extinction and some, such as the disturbed Tokios coral reef formations in the Indian Ocean, are effectively lost. In October 2008, Peter Brabeck-Letmathe, chairman and former chief executive of Nestlé, warned that the production of biofuels will further deplete the world's water supply.

Overview of Regions Suffering Crisis Impacts

There are many other countries of the world that are severely impacted with regard to human health and inadequate drinking water. The following is a partial list of some of the countries with significant populations (numerical population of affected population listed) whose only consumption is of contaminated water:

- Sudan 12.3 million

- Venezuela 5.0 million

- Ethiopia 2.7 million

- Tunisia 2.1 million

- Cuba 1.3 million

Several world maps showing various aspects of the problem can be found in this graph article.

According to the California Department of Resources, if more supplies aren't found by 2020, the region will face a shortfall nearly as great as the amount consumed today. Los Angeles is a coastal desert able to support at most 1 million people on its own water; the Los Angeles basin now is the core of a megacity that spans 220 miles (350 km) from Santa Barbara to the Mexican border. The region's population is expected to reach 41 million by 2020, up from 28 million in 2009. The population of California continues to grow by more than two million a year and is expected to reach 75 million in 2030, up from 49 million in 2009. But water shortage is likely to surface well before then.

Water deficits, which are already spurring heavy grain imports in numerous smaller countries, may soon do the same in larger countries, such as China and India. The water tables are falling in scores of countries (including Northern China, the US, and India) due to widespread overpumping using powerful diesel and electric pumps. Other countries affected include Pakistan, Iran, and Mexico. This will eventually lead to water scarcity and cutbacks in grain harvest. Even with the overpumping of its aquifers, China is developing a grain deficit. When this happens, it will almost certainly drive grain prices upward. Most of the 3 billion people projected to be added worldwide by mid-century will be born in countries already experiencing water shortages. Unless population growth can be slowed quickly, it is feared that there may not be a practical non-violent or humane solution to the emerging world water shortage.

After China and India, there is a second tier of smaller countries with large water deficits — Algeria, Egypt, Iran, Mexico, and Pakistan. Four of these already import a large share of their grain. But with a population expanding by 4 million a year, they will also likely soon turn to the world market for grain.

According to a UN climate report, the Himalayan glaciers that are the sources of Asia's biggest rivers – Ganges, Indus, Brahmaputra, Yangtze, Mekong, Salween and Yellow – could disappear by 2035 as temperatures rise. It was later revealed that the source used by the UN climate report actually stated 2350, not 2035. Approximately 2.4 billion people live in the drainage basin of the Himalayan rivers. India, China, Pakistan, Bangladesh, Nepal and Myanmar could experience floods followed by droughts in coming decades. In India alone, the Ganges provides water for drinking and farming for more than 500 million people. The west coast of North America, which gets much of its water from glaciers in mountain ranges such as the Rocky Mountains and Sierra Nevada, also would be affected.

By far the largest part of Australia is desert or semi-arid lands commonly known as the outback. In June 2008 it became known that an expert panel had warned of long term, possibly irreversible, severe ecological damage for the whole Murray-Darling basin if it does not receive sufficient water by October. Water restrictions are currently in place in many regions and cities of Australia in response to chronic shortages resulting from drought. The Australian of the year 2007, environmentalist Tim Flannery, predicted that unless it made drastic changes, Perth in Western Australia could become the world's first ghost metropolis, an abandoned city with no more water to sustain its population. However, Western Australia's dams reached 50% capacity for the first time since 2000 as of September 2009. As a result, heavy rains have brought forth positive results for the region. Nonetheless, the following year, 2010, Perth suffered its second-driest winter on record and the water corporation tightened water restrictions for spring.

Physical and Economic Scarcity

Around one fifth of the world's population currently live in regions affected by Physical water scarcity, where there is inadequate water resources to meet a country's or regional demand, including the water needed to fulfill the demand of ecosystems to function effectively. Arid regions frequently suffer from physical water scarcity. It also occurs where water seems abundant but where resources are over-committed, such as when there is over development of hydraulic infrastructure for irrigation. Symptoms of physical water scarcity include environmental degradation and declining groundwater as well as other forms of exploitation or overuse.

Economic water scarcity is caused by a lack of investment in infrastructure or technology to draw water from rivers, aquifers or other water sources, or insufficient human capacity to satisfy the demand for water. One quarter of the world's population is affected by economic water scarcity. Economic water scarcity includes a lack of infrastructure, causing the people without reliable access to water to have to travel long distances to fetch water, that is often contaminated from rivers for domestic and agricultural uses. Large parts of Africa suffer from economic water scarcity; developing water infrastructure in those areas could therefore help to reduce poverty. Critical conditions often arise for economically poor and politically weak communities living in already dry environment. Consumption increases with GDP per capita in most developed countries the average amount is around 200-300 litres daily. In underdeveloped (e.g. African countries such as Mozambique, average daily water consumption per capita was below 10 l. This is against the backdrop of international organisations, which recommend a minimum of 20 l of water (not including the water needed for washing clothes), available at most 1 km from the household. Increased water consumption is correlated with increasing income, as measured by GDP per capita. In countries suffering from water shortages water is the subject of speculation.

Human Right to Water

The United Nations Committee on Economic, Social and Cultural Rights established a foundation of five core attributes for water security. They declare that the human right to water entitles everyone to sufficient, safe, acceptable, physically accessible, and affordable water for personal and domestic use.

Millennium Development Goals (MDG)

At the 2000 Millennium Summit, the United Nations addressed the effects of economic water scarcity by making increased access to safe drinking water an international development goal. During this time, they drafted the Millennium Development Goals and all 189 UN members agreed on eight goals. MDG 7 sets a target for reducing the proportion of the population without sustainable safe drinking water access by half by 2015. This would mean that more than 600 million people would gain access to a safe source of drinking water. In 2016, the Sustainable Development Goals replace the Millennium Development Goals.

Water Scarcity's Effects on Environment

An abandoned ship in the former Aral Sea, near Aral, Kazakhstan.

Water scarcity has many negative impacts on the environment, including lakes, rivers, wetlands, and other fresh water resources. The resulting water overuse that is related to water scarcity, often located in areas of irrigation agriculture, harms the environment in several ways including increased salinity, nutrient pollution, and the loss of floodplains and wetlands. Furthermore, water scarcity makes flow management in the rehabilitation of urban streams problematic.

Through the last hundred years, more than half of the Earth's wetlands have been destroyed and have disappeared. These wetlands are important not only because they are the habitats of numerous inhabitants such as mammals, birds, fish, amphibians, and invertebrates, but they support the growing of rice and other food crops as well as provide water filtration and protection from storms and flooding. Freshwater lakes such as the Aral Sea in central Asia have also suffered. Once the fourth largest freshwater lake, it has lost more than 58,000 square km of area and vastly increased in salt concentration over the span of three decades.

Subsidence, or the gradual sinking of landforms, is another result of water scarcity. The U.S. Geological Survey estimates that subsidence has affected more than 17,000 square miles in 45 U.S. states, 80 percent of it due to groundwater usage. In some areas east of Houston, Texas the land has dropped by more than nine feet due to subsidence. Brownwood, a subdivision near Baytown, Texas, was abandoned due to frequent flooding caused by subsidence and has since become part of the Baytown Nature Center.

Climate Change

Aquifer drawdown or overdrafting and the pumping of fossil water increases the total amount of water within the hydrosphere subject to transpiration and evaporation processes, thereby causing accretion in water vapour and cloud cover, the primary absorbers of infrared radiation in the earth's atmosphere. Adding water to the system has a forcing effect on the whole earth system, an accurate estimate of which hydrogeological fact is yet to be quantified.

Depletion of Freshwater Resources

Apart from the conventional surface water sources of freshwater such as rivers and lakes, other resources of freshwater such as groundwater and glaciers have become more developed sources of freshwater, becoming the main source of clean water. Groundwater is water that has pooled below the surface of the Earth and can provide a usable quantity of water through springs or wells. These areas where groundwater is collected are also known as aquifers. Glaciers provide freshwater in the form meltwater, or freshwater melted from snow or ice, that supply streams or springs as temperatures rise. More and more of these sources are being drawn upon as conventional sources' usability decreases due to factors such as pollution or disappearance due to climate changes. The exponential growth rate of the human population is a main contributing factor in the increasing use of these types of water resources.

Groundwater

Until recent history, groundwater was not a highly utilized resource. In the 1960s, more and more groundwater aquifers developed. Changes in knowledge, technology and funding have allowed for focused development into abstracting water from groundwater resources away from surface water resources. These changes allowed for progress in society such as the "agricultural groundwater revolution," expanding the irrigation sector allowing for increased food production and develop-ment in rural areas. Groundwater supplies nearly half of all drinking water in the world. The large volumes of water stored underground in most aquifers have a considerable buffer capacity allow-ing for water to be withdrawn during periods of drought or little rainfall. This is crucial for people that live in regions that cannot depend on precipitation or surface water as a supply alone, instead providing reliable access to water all year round. As of 2010, the world's aggregated groundwater abstraction is estimated at approximately 1,000 km³ per year, with 67% used for irrigation, 22% used for domestic purposes and 11% used for industrial purposes. The top ten major consumers of abstracted water (India, China, United States of America, Pakistan, Iran, Bangladesh, Mexico, Saudi Arabia, Indonesia, and Italy) make up 72% of all abstracted water use worldwide. Ground-water has become crucial for the livelihoods and food security of 1.2 to 1.5 billion rural households in the poorer regions of Africa and Asia.

Although groundwater sources are quite prevalent, one major area of concern is the renewal rate or recharge rate of some groundwater sources. Abstracting from groundwater sources that are non-renewable could lead to exhaustion if not properly monitored and managed. Another concern of increased groundwater usage is the diminished water quality of the source over time. Reduction of natural outflows, decreasing stored volumes, declining water levels and water degradation are commonly observed in groundwater systems. Groundwater depletion may result in many neg-ative effects such as increased cost of groundwater pumping, induced salinity and other water quality changes, land subsidence, degraded springs and reduced baseflows. Human pollution is also harmful to this important resource.

Glaciers

Glaciers are noted as a vital water source due to their contribution to stream flow. Rising global temperatures have noticeable effects on the rate at which glaciers melt, causing glaciers in general to shrink worldwide. Although the meltwater from these glaciers are increasing the total water supply for the present, the disappearance of glaciers in the long term will diminish available water resources. Increased meltwater due to rising global temperatures can also have negative effects such as flooding of lakes and dams and catastrophic results.

Measurement of Water Scarcity

In 2012 in Sindh, Pakistan a shortage of clean water led people to queue to collect it where available

Hydrologists today typically assess water scarcity by looking at the population-water equation. This is done by comparing the amount of total available water resources per year to the population of a country or region. A popular approach to measuring water scarcity has been to rank countries according to the amount of annual water resources available per person. For example, according to the Falkenmark Water Stress Indicator, a country or region is said to experience "water stress" when annual water supplies drop below 1,700 cubic metres per person per year. At levels between 1,700 and 1,000 cubic metres per person per year, periodic or limited water shortages can be expected. When water supplies drop below 1,000 cubic metres per person per year, the country faces "water scarcity". The United Nations' FAO states that by 2025, 1.9 billion people will live in countries or regions with absolute water scarcity, and two-thirds of the world population could be under stress conditions. The World Bank adds that climate change could profoundly alter future patterns of both water availability and use,thereby increasing levels of water stress and insecurity, both at the global scale and in sectors that depend on water.

Other ways of measuring water scarcity include examining the physical existence of water in nature, comparing nations with lower or higher volumes of water available for use. This method often fails to capture the accessibility of the water resource to the population that may need it. Others have related water availability to population.

Another measurement, calculated as part of a wider assessment of water management in 2007, aimed to relate water availability to how the resource was actually used. It therefore divided water scarcity into 'physical' and 'economic'. Physical water scarcity is where there is not enough water to meet all demands, including that needed for ecosystems to function effectively. Arid regions frequently suffer from physical water scarcity. It also occurs where water seems abundant but where resources are over-committed, such as when there is overdevelopment of hydraulic infrastructure for irrigation. Symptoms of physical water scarcity include environmental degradation and declining groundwater. Water stress harms living things because every organism needs water to live.

Renewable Freshwater Resources

Renewable freshwater supply is a metric often used in conjunction when evaluating water scarcity. This metric is informative because it can describe the total available water resource each country contains. By knowing the total available water source, an idea can be gained about whether a country is prone to experiencing physical water scarcity. This metric has its faults in that it is an average; precipitation delivers water unevenly across the planet each year and annual renewable water resources vary from year to year. This metric also does not describe the accessibility of water to individuals, households, industries, or the government. Lastly, as this metric is a description of a whole country, it does not accurately portray whether a country is experiencing water scarcity. Canada and Brazil both have very high levels of available water supply, but still experience various water related problems.

It can be observed that tropical countries in Asia and Africa have low availability of freshwater resources.

The following table displays the average annual renewable freshwater supply by country including both surface-water and groundwater supplies. This table represents data from the UN FAO AQUASTAT, much of which are produced by modeling or estimation as opposed to actual measurements.

Total renewable freshwater supply by country				
Rank	Country	Annual renewable water resources (km³/year)	Region	Year of estimate
1	Kuwait	0.02	Asia	2008
2	St. Kitts and Nevis	0.02	North and Central America	2000
3	Maldives	0.03	Asia	1999
4	Malta	0.07	Europe	2005
5	Antigua and Barbuda	0.1	North and Central America	2000
6	Qatar	0.1	Asia	2008
7	Barbados	0.1	North and Central America	2003
8	Bahrain	0.1	Asia	2008
9	United Arab Emirates	0.2	Asia	2008
10	Cape Verde	0.3	Africa	2005
11	Djibouti	0.3	Africa	2005

12	Cyprus	0.3	Europe	2007
13	Libya	0.6	Africa	2005
14	Singapore	0.6	Asia	1975
15	Jordan	0.9	Asia	2008
16	Comoros	1.2	Africa	2005
17	Oman	1.4	Asia	2008
18	Luxembourg	1.6	Europe	2007
19	Israel	1.8	Asia	2008
20	Yemen	2.1	Asia	2008
21	Saudi Arabia	2.4	Asia	2008
22	Mauritius	2.8	Africa	2005
23	Burundi	3.6	Africa	1987
24	Trinidad and Tobago	3.8	North and Central America	2000
25	Swaziland	4.5	Africa	1987
26	Lebanon	4.5	Asia	2008
27	Tunisia	4.6	Africa	2005
28	Reunion	5.0	Africa	1988
29	Lesotho	5.2	Africa	1987
30	Eritrea	6.3	Africa	2001
31	Macedonia	6.4	Europe	2001
32	Armenia	7.8	Former Soviet Union	2008
33	Gambia	8.0	Africa	2005
34	Brunei	8.5	Asia	1999
35	Jamaica	9.4	North and Central America	2000
36	Rwanda	9.5	Africa	2005
37	Mauritania	11.4	Africa	2005
38	Algeria	11.6	Africa	2005
39	Moldova	11.7	Former Soviet Union	1997
40	Estonia	12.3	Europe	2007
41	Estonia	12.8	Former Soviet Union	1997
42	Haiti	14.0	North and Central America	2000
43	Somalia	14.2	Africa	2005
44	Botswana	14.7	Africa	2001
45	Togo	14.7	Africa	2001
46	Czech Republic	16.0	Europe	2007
47	Denmark	16.3	Europe	2007
48	Syria	16.8	Asia	2008
49	Malawi	17.3	Africa	2001

50	Burkina Faso	17.5	Africa	2001
51	Namibia	17.7	Africa	2005
52	Belize	18.6	North and Central America	2000
53	Zimbabwe	20.0	Africa	1987
54	Belgium	20.0	Europe	2007
55	Dominican Republic	21.0	North and Central America	2000
56	Lithuania	24.5	Former Soviet Union	2007
57	El Salvador	25.2	North and Central America	2001
58	Romania	25.7	Europe	2007
59	Benin	25.8	Africa	2001
60	Equatorial Guinea	26	Africa	2001
61	Fiji	28.6	Oceania	1987
62	Morocco	29.0	Africa	2005
63	Kenya	30.7	Africa	2005
64	Guinea-Bissau	31.0	Africa	2005
65	Slovenia	32.1	Europe	2007
66	Niger	33.7	Africa	2005
67	Azerbaijan	34.7	Former Soviet Union	2008
68	Mongolia	34.8	Asia	1999
69	Bosnia and Herzegovina	37.5	Europe	2003
70	Cuba	38.1	North and Central America	2000
71	Senegal	39.4	Africa	1987
72	Albania	41.7	Europe	2001
73	Chad	43.0	Africa	1987
74	Solomon Islands	44.7	Oceania	1987
75	Kyrgyzstan	46.5	Former Soviet Union	1997
76	Ireland	46.8	Europe	2003
77	South Africa	50.0	Africa	2005
78	Sri Lanka	50.0	Asia	1999
79	Slovakia	50.1	Europe	2007
80	Ghana	53.2	Africa	2001
81	Switzerland	53.5	Europe	2007
82	Belarus	58.0	Former Soviet Union	1997
83	Egypt	58.3	Africa	2005
84	Turkmenistan	60.9	Former Soviet Union	1997
85	Poland	63.1	Europe	2007

86	Georgia	63.3	Former Soviet Union	2008
87	Sudan	64.5	Africa	2005
88	Afghanistan	65.0	Asia	1997
89	Uganda	66.0	Africa	2005
90	Taiwan	67.0	Asia	2000
91	Korea Rep	69.7	Asia	1999
92	Greece	72.0	Europe	2007
93	Uzbekistan	72.2	Former Soviet Union	2003
94	Portugal	73.6	Europe	2007
95	Iraq	75.6	Asia	2008
96	Korea DPR	77.1	Asia	1999
97	Côte d'Ivoire	81	Africa	2001
98	Austria	84.0	Europe	2007
99	Netherlands	89.7	Europe	2007
100	Tanzania	91	Africa	2001
101	Bhutan	95.0	Asia	1987
102	Honduras	95.9	North and Central America	2000
103	Tajikistan	99.7	Former Soviet Union	1997
104	Mali	100.0	Africa	2005
105	Zambia	105.2	Africa	2001
106	Croatia	105.5	Europe	1998
107	Bulgaria	107.2	Europe	2010
108	Kazakhstan	109.6	Former Soviet Union	1997
109	Ethiopia	110.0	Africa	1987
110	Finland	110.0	Europe	2007
111	Spain	111.1	Europe	2007
112	Guatemala	111.3	North and Central America	2000
113	Costa Rica	112.4	North and Central America	2000
114	Hungary	116.4	Europe	2007
115	Suriname	122.0	South America	2003
116	Iran	137.5	Asia	2008
117	Uruguay	139.0	South America	2000
118	Ukraine	139.5	Former Soviet Union	1997
119	Central African Republic	144.4	Africa	2005
120	Panama	148.0	North and Central America	2000
121	Sierra Leone	160.0	Africa	1987

122	Gabon	164.0	Africa	1987
123	Iceland	170.0	Europe	2007
124	Italy	175.0	Europe	2007
125	United Kingdom	175.3	Europe	2007
126	Sweden	183.4	Europe	2007
127	Angola	184.0	Africa	1987
128	France	186.3	Europe	2007
129	Germany	188.0	Europe	2007
130	Nicaragua	196.7	North and Central America	2000
131	Serbia-Montenegro*	208.5	Europe	2003
132	Nepal	210.2	Asia	1999
133	Turkey	213.6	Asia	2008
134	Mozambique	217.1	Africa	2005
135	Guinea	226.0	Africa	1987
136	Liberia	232.0	Africa	1987
137	Pakistan	233.8	Asia	2003
138	Guyana	241.0	South America	2000
139	Cameroon	285.5	Africa	2003
140	Nigeria	286.2	Africa	2005
141	Laos	333.6	Asia	2003
142	Paraguay	336.0	South America	2000
143	Australia	336.1	Oceania	2005
144	Madagascar	337.0	Africa	2005
145	Latvia	337.3	Former Soviet Union	2007
146	Norway	389.4	Europe	2007
147	New Zealand	397.0	Oceania	1995
148	Thailand	409.9	Asia	1999
149	Japan	430.0	Asia	1999
150	Ecuador	432.0	South America	2000
151	Mexico	457.2	North and Central America	2000
152	Cambodia	476.1	Asia	1999
153	Philippines	479.0	Asia	1999
154	Malaysia	580.0	Asia	1999
155	Bolivia	622.5	South America	2000
156	Papua New Guinea	801.0	Oceania	1987
157	Argentina	814.0	South America	2000
158	Congo	832.0	Africa	1987
159	Vietnam	891.2	Asia	1999
160	Chile	922.0	South America	2000
161	Myanmar	1045.6	Asia	1999

162	Bangladesh	1210.6	Asia	1999
163	Venezuela	1233.2	South America	2000
164	Congo, Democratic Republic (formerly Zaire)	1283	Africa	2001
165	India	1907.8	Asia	1999
166	Peru	1913.0	South America	2000
167	Colombia	2132.0	South America	2000
168	China	2738.8	Asia	2008
169	Indonesia	2838.0	Asia	1999
170	United States of America	3069.0	North and Central America	1985
171	Canada	3300.0	North and Central America	1985
172	Russia	4498.0	Former Soviet Union	1997
173	Brazil	8233.0	South America	2000

Outlook

Wind and solar power such as this installation in a village in northwest
Madagascar can make a difference in safe water supply.

Construction of wastewater treatment plants and reduction of groundwater overdrafting appear to
be obvious solutions to the worldwide problem; however, a deeper look reveals more fundamental
issues in play. Wastewater treatment is highly capital intensive, restricting access to this technology in some regions; furthermore the rapid increase in population of many countries makes this
a race that is difficult to win. As if those factors are not daunting enough, one must consider the
enormous costs and skill sets involved to maintain wastewater treatment plants even if they are
successfully developed.

Reducing groundwater overdrafting is usually politically unpopular, and can have major economic
impacts on farmers. Moreover, this strategy necessarily reduces crop output, something the world
can ill-afford given the current population.

At more realistic levels, developing countries can strive to achieve primary wastewater treatment
or secure septic systems, and carefully analyse wastewater outfall design to minimise impacts to

drinking water and to ecosystems. Developed countries can not only share technology better, including cost-effective wastewater and water treatment systems but also in hydrological transport modeling. At the individual level, people in developed countries can look inward and reduce over-consumption, which further strains worldwide water consumption. Both developed and developing countries can increase protection of ecosystems, especially wetlands and riparian zones. These measures will not only conserve biota, but also render more effective the natural water cycle flushing and transport that make water systems more healthy for humans.

A range of local, low-tech solutions are being pursued by a number of companies. These efforts center around the use of solar power to distill water at temperatures slightly beneath that at which water boils. By developing the capability to purify any available water source, local business models could be built around the new technologies, accelerating their uptake. For example, bedouins from the town of Dahab in Egypt have installed AquaDania's WaterStillar, which uses a solar thermal collector measuring two square metres to distill from 40 to 60 litres per day from any local water source. This is five times more efficient than conventional stills and eliminates the need for polluting plastic PET bottles or transportation of water supply.

Global Experiences in Managing Water Crisis

It is alleged that the likelihood of conflict rises if the rate of change within the basin exceeds the capacity of institution to absorb that change. Although water crisis is closely related to regional tensions, history showed that acute conflicts over water are far less than the record of cooperation.

The key lies in strong institutions and cooperation. The Indus River Commission and the Indus Water Treaty survived two wars between India and Pakistan despite their hostility, proving to be a successful mechanism in resolving conflicts by providing a framework for consultation inspection and exchange of data. The Mekong Committee has also functioned since 1957 and survived the Vietnam War. In contrast, regional instability results when there is an absence of institutions to co-operate in regional collaboration, like Egypt's plan for a high dam on the Nile. However, there is currently no global institution in place for the management and management of trans-boundary water sources, and international co-operation has happened through ad hoc collaborations between agencies, like the Mekong Committee which was formed due to an alliance between UNICEF and the US Bureau of Reclamation. Formation of strong international institutions seems to be a way forward – they fuel early intervention and management, preventing the costly dispute resolution process.

One common feature of almost all resolved disputes is that the negotiations had a "need-based" instead of a "right–based" paradigm. Irrigable lands, population, technicalities of projects define "needs". The success of a need-based paradigm is reflected in the only water agreement ever negotiated in the Jordan River Basin, which focuses in needs not on rights of riparians. In the Indian subcontinent, irrigation requirements of Bangladesh determine water allocations of The Ganges River. A need based, regional approach focuses on satisfying individuals with their need of water, ensuring that minimum quantitative needs are being met. It removes the conflict that arises when countries view the treaty from a national interest point of view, move away from the zero-sum approach to a positive sum, integrative approach that equitably allocated the water and its benefits.

The Blue Peace framework developed by Strategic Foresight Group in partnership with the Governments of Switzerland and Sweden offers a unique policy structure which promotes sustainable management of water resources combined with cooperation for peace. By making the most of shared water resources through cooperation rather than mere allocation between countries, the chances for peace can be increased. The Blue Peace approach has proven to be effective in cases like the Middle East and the Nile basin.

References

- Environment Canada (2005). Municipal Water Use, 2001 Statistics (PDF) (Report). Retrieved 2010-02-02. Cat. No. En11-2/2001E-PDF. ISBN 0-662-39504-2. p. 3.

- Prokurat, Sergiusz (2015), Drought and water shortages in Asia as a threat and economic problem (PDF), Józefów: "Journal of Modern Science" 3/26/2015, pp. 235–250, retrieved 5 August 2016

- Santos, Jessica; van der Linden, Sander (2016). "Changing Norms by Changing Behavior: The Princeton Drink Local Program". Environmental Practice. 18 (2): 1–7. doi:10.1017/S1466046616000144.

- Brown, Lester R. (8 September 2002) Water Shortages May Cause Food Shortages. Greatlakesdirectory.org. Retrieved on 27 August 2013.

- WWAP (World Water Assessment Programme). 2012. The United Nations World Water Development Report 4: Managing Water under Uncertainty and Risk. Paris, UNESCO.

- The World Bank, 2009 "Water and Climate Change: Understanding the Risks and Making Climate-Smart Investment Decisions". pp. 21–24. Retrieved 24 October 2011.

- Brown, Lester R. (27 September 2006). "Water Scarcity Crossing National Borders". Earth Policy Institute. Archived from the original on 2009-03-31. Retrieved 10 March 2011.

- U.S. Environmental Protection Agency (EPA) (2002). Cases in Water Conservation (PDF) (Report). Retrieved 2010-02-02. Document No. EPA-832-B-02-003.

- Albuquerque Bernalillo County Water Utility Authority (2009-02-06). "Xeriscape Rebates". Albuquerque, NM. Retrieved 2010-02-02.

8

Conflicts on Water Resources

Water conflict is a term describing a conflict between countries, states or groups over water resources. Water has historically been a source of tension and a factor of conflicts; this chapter elaborates this aspect on water. The following chapter will provide an integrated understanding of water resources.

Water Conflict

Water conflict is a term describing a conflict between countries, states, or groups over an access to water resources. The United Nations recognizes that water disputes result from opposing interests of water users, public or private.

A wide range of water conflicts appear throughout history, though rarely are traditional wars waged over water alone. Instead, water has historically been a source of tension and a factor in conflicts that start for other reasons. However, water conflicts arise for several reasons, including territorial disputes, a fight for resources, and strategic advantage. A comprehensive online database of water-related conflicts—the Water Conflict Chronology—has been developed by the Pacific Institute. This database lists violence over water going back nearly 5,000 years.

These conflicts occur over both freshwater and saltwater, and both between and within nations. However, conflicts occur mostly over freshwater; because freshwater resources are necessary, yet limited, they are the center of water disputes arising out of need for potable water and irrigation. As freshwater is a vital, yet unevenly distributed natural resource, its availability often impacts the living and economic conditions of a country or region. The lack of cost-effective water supply options in areas like the Middle East, among other elements of water crises can put severe pressures on all water users, whether corporate, government, or individual, leading to tension, and possibly aggression. Recent humanitarian catastrophes, such as the Rwandan Genocide or the war in Sudanese Darfur, have been linked back to water conflicts.

A recent report "Water Cooperation for a Secure World" published by Strategic Foresight Group concludes that active water cooperation between countries reduces the risk of war. This conclusion is reached after examining trans-boundary water relations in over 200 shared river basins in 148 countries, though as noted below, a growing number of water conflicts are sub-national.

Causes

According to the 1992 International Conference on Water and the Environment, water is a vital element for human life, and human activities are closely connected to availability and quality of water. Unfortunately, water is a limited resource and in the future access "might get worse with climate change, although scientists' projections of future rainfall are notoriously cloudy" writes Roger Harrabin. Moreover, "it is now commonly said that future wars in the Middle East are more likely to be fought over water than over oil," said Lester R. Brown at a previous Stockholm Water Conference.

Water conflicts occur because the demand for water resources and potable water can exceed supply, or because control over access and allocation of water may be disputed. Elements of a water crisis may put pressures on affected parties to obtain more of a shared water resource, causing diplomatic tension or outright conflict.

11% of the global population, or 783 million people, are still without access to improved sources of drinking water which provides the catalyst for potential for water disputes. Besides life, water is necessary for proper sanitation, commercial services, and the production of commercial goods. Thus numerous types of parties can become implicated in a water dispute. For example, corporate entities may pollute water resources shared by a community, or governments may argue over who gets access to a river used as an international or inter-state boundary.

The broad spectrum of water disputes makes them difficult to address. Locale, local and international law, commercial interests, environmental concerns, and human rights questions make water disputes complicated to solve – combined with the sheer number of potential parties, a single dispute can leave a large list of demands to be met by courts and lawmakers.

Economic and Trade Issues

Water's viability as a commercial resource, which includes fishing, agriculture, manufacturing, recreation and tourism, among other possibilities, can create dispute even when access to potable water is not necessarily an issue. As a resource, some consider water to be as valuable as oil, needed by nearly every industry, and needed nearly every day. Water shortages can completely cripple an industry just as it can cripple a population, and affect developed countries just as they affect countries with less-developed water infrastructure. Water-based industries are more visible in water disputes, but commerce at all levels can be damaged by a lack of water.

International commercial disputes between nations can be addressed through the World Trade Organization, which has water-specific groups like a Fisheries Center that provide a unified judicial protocol for commercial conflict resolution. Still, water conflict occurring domestically, as well as conflict that may not be entirely commercial in nature may not be suitable for arbitration by the WTO.

Fishing

Historically, like fisheries have been the main sources of question, as nations expanded and claimed portions of oceans and seas as territory for 'domestic' commercial fishing. Certain lucrative areas, such as the Bering Sea, have a history of dispute; in 1886 Great Britain and the United States clashed over sealing fisheries, and today Russia surrounds a pocket of international water known as the Bering Sea Donut Hole. Conflict over fishing routes and access to the hole was resolved in 1995 by a convention referred to colloquially as the Donut Hole Agreement.

Pollution

Corporate interest often crosses opposing commercial interest, as well as environmental concerns, leading to another form of dispute. In the 1960s, Lake Erie, and to a lesser extent, the other Great Lakes were polluted to the point of massive fish death. Local communities suffered greatly from dismal water quality until the United States Congress passed the Clean Water Act in 1972.

Water pollution poses a significant health risk, especially in heavily industrialized, heavily populated areas like China. In response to a worsening situation in which entire cities lacked safe drinking water, China passed a revised Water Pollution Prevention and Control Law. The possibility of polluted water making it way across international boundaries, as well as unrecognized water pollution within a poorer country brings up questions of human rights, allowing for international input on water pollution. There is no single framework for dealing with pollution disputes local to a nation.

Classifications

According to Aaron Wolf, et all. there were 1831 water conflicts over transboundary basins from 1950–2000. They categorized these events as following:

- No water-related events on the extremes

- Most interactions are cooperative

- Most interactions are mild

- Water acts as irritant

- Water acts as unifier

- Nations cooperate over a wide variety of issues

- Nations conflict over quantity and infrastructure

A comprehensive chronology of water-related conflicts is maintained by the Pacific Institute in their Water Conflict Chronology, which includes an open-source data set, an interactive map, and full information on citations. These historical examples go back over 4,500 years. In this dataset, water conflicts are categorized as follows:

- Control of Water Resources (state and non-state actors): where water supplies or access to water is at the root of tensions.

- Military Tool (state actors): where water resources, or water systems themselves, are used by a nation or state as a weapon during a military action.

- Political Tool (state and non-state actors): where water resources, or water systems themselves, are used by a nation, state, or non-state actor for a political goal.

- Terrorism (non-state actors): where water resources, or water systems, are either targets or tools of violence or coercion by non-state actors.

- Military Target (state actors): where water resource systems are targets of military actions by nations or states.

- Development Disputes (state and non-state actors): where water resources or water systems are a major source of contention and dispute in the context of economic and social development

Response

International organizations play the largest role in mediating water disputes and improving water management. From scientific efforts to quantify water pollution, to the World Trade Organization's efforts to resolve trade disputes between nations, the varying types of water disputes can be addressed through current framework. Yet water conflicts that go unresolved become more dangerous as water becomes more scarce and global population increases.

United Nations

The UN International Hydrological Program aims to help improve understanding of water resources and foster effective water management. But by far the most active UN program in water dispute resolution is its Potential Conflict to Co-operation Potential (PCCP) mission, which is in its third phase, training water professionals in the Middle East and organizing educational efforts elsewhere. Its target groups include diplomats, lawmakers, civil society, and students of water studies; by expanding knowledge of water disputes, it hopes to encourage co-operation between nations in dealing with conflicts.

UNESCO has published a map of transboundary aquifers. Academic work focusing on water disputes has yet to yield a consistent method for mediating international disputes, let alone local ones. But UNESCO faces optimistic prospects for the future as water conflicts become more public, and as increasing severity sobers obstinate interests.

World Trade Organization

The World Trade Organization can arbitrate water disputes presented by its member states when the disputes are commercial in nature. The WTO has certain groups, such as its Fisheries Center, that work to monitor and rule on relevant cases, although it is by no means the authority on conflict over water resources.

Because water is so central to agricultural trade, water disputes may be subtly implicated in WTO cases in the form of virtual water, water used in the production of goods and services but not directly traded between countries. Countries with greater access to water supplies may fare better from an economic standpoint than those facing crisis, which creates the potential for conflict. Outraged by agriculture subsidies that displace domestic produce, countries facing water shortages bring their case to the WTO.

The WTO plays more of a role in agriculturally based disputes that are relevant to conflict over specific sources of water. Still, it provides an important framework that shapes the way water will play into future economic disputes. One school of thought entertains the notion of war over water, the ultimate progression of an unresolved water dispute—scarce water resources combined with the pressure of exponentially increasing population may outstrip the ability of the WTO to maintain civility in trade issues

Notable Conflicts

Water conflicts can occur on the intrastate and interstate levels. Interstate conflicts occur between two or more neighboring countries that share a transboundary water source, such as a river, sea, or groundwater basin. For example, the Middle East has only 1% of the world's freshwater shared among 5% of the world's population. Intrastate conflicts take place between two or more parties in the same country. An example would be the conflicts between farmers and industry (agricultural vs industrial use of water).

According to UNESCO, the current interstate conflicts occur mainly in the Middle East (disputes stemming from the Euphrates and Tigris Rivers among Turkey, Syria, and Iraq; and the Jordan River conflict among Israel, Lebanon, Jordan and the State of Palestine), in Africa (Nile River-related conflicts among Egypt, Ethiopia, and Sudan), as well as in Central Asia (the Aral Sea conflict among Kazakhstan, Uzbekistan, Turkmenistan, Tajikistan and Kyrgyzstan). At a local level, a remarkable example is the 2000 Cochabamba protests in Bolivia, depicted in the 2010 Spanish film *Even the Rain* by Icíar Bollaín.

Some analysts estimate that due to an increase in human consumption of water resources, water conflicts will become increasingly common in the near future.

In 1979 Egyptian President Anwar Sadat said that if Egypt were to ever go to war again it would be over water. Separately, amidst Egypt–Ethiopia relations, Ethiopian Prime Minister Meles Zenawi said: "I am not worried that the Egyptians will suddenly invade Ethiopia. Nobody who has tried that has lived to tell the story."

List of Notable Water Conflicts in India

1. Brahmaputra
2. Godavari
3. Ganga
4. Hirakud
5. Kaveri
6. Narmada

Recent Research into Water Conflicts

Some research from the International Water Management Institute and Oregon State University has found that water conflicts among nations are less likely than is cooperation, with hundreds of treaties and agreements in place. Water conflicts tend to arise as an outcome of other social issues. Conversely, the Pacific Institute has shown that while interstate (i.e., nation to nation) water conflicts are increasingly less likely, there appears to be a growing risk of sub-national conflicts among water users, regions, ethnic groups, and competing economic interests. Data from the Water Conflict Chronology show these intrastate conflicts to be a larger and growing component of all water disputes, and that the traditional international mechanisms for addressing them, such as bilateral or multilateral treaties, are not as effective.

Strategic Foresight Group in partnership with the Governments of Switzerland and Sweden has developed the Blue Peace approach which seeks to transforms trans-boundary water issues into instruments for cooperation. The Blue Peace framework offers a unique policy structure which promotes sustainable management of water resources combined with cooperation for peace. By making the most of shared water resources through cooperation rather than mere allocation between countries, the chances for peace can be increased. The Blue Peace approach has proven to be effective in cases like the Middle East and the Nile basin.

References

- Hassan, Fekri A. (2003), Water Management and Early Civilizations: From Cooperation to Conflict (PDF), UNESCO, retrieved 1 May 2015

- Promoting cooperation through management of trans-boundary water resources, Success Stories, Issue 8, 2010, IWMI.

9

Shared Vision Planning: A Different Approach

Planning is an important component of any field of study. Shared vision planning seeks to be an update on existing water resource management systems. Decisions are made through a process of "informed consent." The content serves as a source to understand the process of shared vision planning.

Shared vision planning was developed by the U.S. Army Corps of Engineers during the National Drought Study (1989-1993). Shared vision planning has three basic elements: (1) an updated version of the systems approach to water resources management developed during the Harvard Water Program; (2) an approach to public involvement called "Circles of Influence"; and (3) collaboratively built computer models of the system to be managed. Alternative dispute resolution methods are often used to bring people in conflict to the table, and to resolve differences that occur during planning. A method of collaborative decision making called "Informed consent" is used to make decisions internally consistent, more defensible and transparent.

Three Basic Elements

Each of the three elements in shared vision planning has an evolutionary history leading to it. The Harvard Water Program planning approach (documented in a 1962 book, "Design of Water Resources Systems; New Techniques for Relating Economic Objectives, Engineering Analysis, and Governmental Planning") is the conceptual starting point for the first element, the planning method. Harvard's conceptual ideas were practiced and refined by a team led by Harry Schwarz in the North Atlantic Regional Study, a Federal water study instigated by President Lyndon B. Johnson after an historic drought in the Northeast led New York City to stop releasing water from its reservoirs into the Delaware River. President Johnson intervened because without that rush of freshwater flowing down the Delaware into the Atlantic Ocean, Atlantic salt water might have been introduced into the Philadelphia drinking water system. Schwarz later summarized the practical interpretation of the Harvard principles in "Large Scale Regional Water Resources Planning" which he wrote with David Major, a Harvard economist. The lessons learned in the North Atlantic Study were later encoded in the Federal "Principles and Standards for Planning Water and Related Land Resources" (1973) often shortened to "the P&S". The "P&S" were used in the design and justification of proposed federal water projects. Eugene Stakhiv, then head of the Corps policy studies, first suggested that a variation on the P&S should be used for drought response planning. During the National Drought Study, the Corps modified the P&S planning steps to make them more suitable for drought management decisions, which generally do not involve significant federal funding. These decisions typically must also be agreed to by multiple management entities, and there are typically multiple levels of government that control the drought response.

The "Circles of Influence" method of public involvement was developed by Dr. Robert Waldman during the National Drought Study and modified traditional Federal approaches. Typically, Federal water studies would invite "stakeholders" - people whose lives could be affected by the decision to build a project - to public meetings. Stakeholders might come to support and shape the infusion of Federal money into the region. But drought management studies typically do not determine whether federal money will be invested in a region; they determine water use curtailment and the storage and allocation of water supplies. Accordingly, "Circles of Influence" management agencies participate in forums already being used by stakeholders, such as city water advisory boards or boating groups. Trust is developed in concentric circles; the planner works to deserve the trust of the leaders other stakeholders already trust. Disputes that arise during planning are addressed using a range of alternative dispute resolution (ADR) methods. The Corps had been an early adapter of ADR, thanks to the work of Jerry Delli Priscoli. Delli Priscoli who worked at the Corps' Institute for Water Resources, where the National Drought Study was managed, was an informal advisor on the Drought Study. While there is still a link between the SVP community and the ADR community, it is largely personal, as the two communities are not served by a single professional society or journal.

The third element, the "shared vision model" was proposed by Richard Palmer during a National Drought Study case study meeting in Seattle, Washington in 1991. More than ten years earlier, Palmer had been doing post-doctorate work from Johns Hopkins University at the Interstate Commission on the Potomac River Basin (ICPRB). ICPRB was one of several government organizations developing reservoir projects to address water supply needs in the Washington Metropolitan area. Palmer proved using an optimization model that regional water needs could be met with only two new reservoirs if the water supply systems of Maryland, Northern Virginia and Washington were managed collaboratively, but because his model was a black box, his solution was not given a fair hearing. Palmer was aware of the use of interactive water models by Pete Loucks, a professor of engineering at Cornell University and he decided to build a simple interactive simulation model he called "PRISM" (Potomac River Interactive Simulation Model) and used it in a game playing exercise with water utility representatives. The exercise led to an historic agreement in 1981. Dan Sheer, who was working at ICPRB labeled this technique "computer aided negotiation" and has been using it since then in river basin studies around the world. Palmer introduced the concept to the National Drought Study team using a systems simulation model building software called STELLA, and the union of Palmer's modeling technique with the planning and public participation elements resulted in what is now known as shared vision planning.

Name

The method at first was called the "drought preparedness study method" but was renamed "Shared Vision Planning" at the suggestion of the late Brian Mar, a University of Washington colleague of Palmer who had heard the name applied to systems design at nearby Boeing Aerospace. Peter Senge used the shared vision phrase in his 1990 book on systems analysis, "The Fifth Discipline" and this may have been where Boeing learned it.

Other Case Studies

Shared vision planning was used in five test studies during the National Drought Study; two were considered successful, the other three were not. Bill Werick, the National Drought Study leader later developed a set of "triage" questions to ask that would help determine whether shared vision planning would be helpful. Shared vision planning use was expanded beyond drought response planning immediately after the National Drought Study in the Alabama-Coosa-Tallapoosa - Apala-chicola-Chattahoochee-Flint (ACT-ACF) Rivers conflict. Its application led to the first interstate water compact in the southeast, but the compact expired years later when the three governors of Florida, Georgia and Alabama could not agree on terms for its continuation. The method has been used in several case studies since, most recently in the International Joint Commission (IJC) studies of Lake Ontario-St. Lawrence River Study (2000-2006) and the Upper Great Lakes (2007-2012). The method has also been applied in Peru in support of some World Bank and Interamerican Development Bank water projects.

Prospects for Wider use

The use of the methods underlying shared vision planning is now more widespread than the name itself. Robert Costanza, the director of the Gund Institute for Ecological Economics, has used STELLA models in environmental and economic modeling since 1987. The International Environmental Modelling and Software Society (iEMSs), which Dr. Costanza co-founded, has as its goal the development and use of environmental modelling and software tools to advance the science and improve decision making with respect to resource and environmental issues. Many iEMSs members have developed approaches similar to shared vision planning without any apparent awareness of the Corps' work, which is perhaps a testament to its soundness. "Mediated Modeling" describes a systems based participatory modeling approach to environmental issues that is very similar to shared vision planning but has typically been used as a learning tool, especially for stakeholders who might otherwise be left out of public policy dialogues.

The Corps' Institute for Water Resources (IWR) managed the National Drought Study and now manages the Corps' shared vision planning activities. IWR, Sandia National Laboratories, and the United States Center for Environmental Conflict Resolution hosted the first national gathering of shared vision planners in Albuquerque, New Mexico in September 2007. A second conference was held in 2010 in Denver, Colorado. The conferences were part of the Computer Aided Dispute Resolution or CADRe program. Conference participants agreed that the basic methods of shared vision planning were slowly becoming part of the mainstream in water resources management but that the process could be accelerated if a common name and community identity could be applied to all the various permutations on the approach.

Laws and Policies Related to Water Resources

Water resources can best be understood with the major topics listed in the following chapter. The laws and policies related to water resources are dealt with in this chapter. Water quality law, water resources law, water rights and xerochore are some of the topics elaborated in the concerned text.

Water Quality Law

Water quality laws govern the release of pollutants into water resources, including surface water, ground water, and stored drinking water. Some water quality laws, such as drinking water regulations, may be designed solely with reference to human health. Many others, including restrictions on the alteration of the chemical, physical, radiological, and biological characteristics of water resources, may also reflect efforts to protect aquatic ecosystems more broadly. Regulatory efforts may include identifying and categorizing water pollutants, dictating acceptable pollutant concentrations in water resources, and limiting pollutant discharges from effluent sources. Regulatory areas include sewage treatment and disposal, industrial and agricultural waste water management, and control of surface runoff from construction sites and urban environments.

Regulated Waters

The Earth's hydrosphere is ubiquitous, fluid, and complex. Within the water cycle, physical water moves without regard to political boundaries between the Earth's atmosphere, surface, and subsurface, through both natural and man-made channels. Consequently, water quality laws must define the portion of this complex system subject to regulatory control. Regulatory jurisdictions may be coterminous with political boundaries (e.g., certain treaty responsibilities may apply to water pollution in all of Earth's international waters). Other laws may apply only to a subset of waters within a political boundary (e.g., a national law that applies only to navigable surface waters), or to a special class of water (e.g., drinking water resources). Cross-jurisdictional waters may be subject to cross-jurisdictional agreements. Even within jurisdictions, complexities may arise where water flows between subsurface and surface, or saturates land without permanently inundating it (wetlands).

Water Pollutant Classification

Water quality laws must identify the substances and energies which qualify as "water pollution" for purposes of further control. From a regulatory perspective, this requires defining the class(es) of materials that qualify as pollutants, and the activities that transform a material into a pollutant.

For example, the United States Clean Water Act defines "pollution" (i.e., water pollution) very broadly to include any and all "man-made or man-induced alteration of the chemical, physical, biological, and radiological integrity of water." However, the Act defines "pollutants" subject to its control more specifically, as "dredged spoil, solid waste, incinerator residue, filter backwash, sewage, garbage, sewage sludge, munitions, chemical wastes, biological materials, radioactive materials [with certain exceptions], heat, wrecked or discarded equipment, rock, sand, cellar dirt and industrial, municipal, and agricultural waste discharged into water. " This definition begins to define both the classes or types of materials (e.g., solid waste) and energies (e.g., heat) that may constitute water pollution, and indicates the moment at which otherwise useful materials may be transformed into pollution for regulatory purposes: when they are "discharged into water," defined elsewhere as "addition" of the material to regulated waters.

Regulatory regimes may also use definitions to reflect policy decisions, excluding certain classes of materials from the definition of water pollution that would otherwise be considered to constitute water pollution. For example, the U.S. Clean Water Act definition quoted above later excludes sewage discharged from vessels (further information at Regulation of ship pollution in the United States) - meaning that a common and important class of water pollution is, by definition, not considered a pollutant for purposes of the United States' primary water quality law.

Definitional questions have resulted in litigation in the United States regarding whether even water itself may qualify as a "pollutant" (e.g., adding warm water to a stream). The United States Supreme Court addressed these issues most recently in *Los Angeles County Flood Control District v. Natural Resources Defense Council, Inc.*.

Water Quality Standards

Water quality standards are legal standards or requirements governing water quality, that is, the concentrations of water pollutants in some regulated volume of water. Such standards are generally expressed as levels of a specific water pollutants (whether chemical, physical, biological, or radiological) that are deemed acceptable in the water volume, and are generally designed relative to the water's intended use - whether for human consumption, industrial or domestic use, recreation, or as aquatic habitat. Determining appropriate water quality standards generally requires up-to-date scientific data on the health or environmental effects of the pollutant under review. It also may require periodic or continuous monitoring of water quality. Regulatory decisions on water quality standards may also incorporate political considerations, such as the economic costs and benefits of compliance.

As an example, the United States employs water quality standards as part of its regulation of surface water quality under the Clean Water Act. The Clean Water Act Water Quality Standards (WQS) Program begins with U.S. states designating intended uses (e.g., recreation, drinking water, natural habitat) for surface water, after which they develop science-based water quality criteria - including numeric pollutant concentration limits, narrative goals (e.g., free from algae blooms), and narrative biological criteria (i.e., the aquatic life that should be able to live in the waterbody). If the water body fails the existing WQS criteria, the state must develop a Total Maximum Daily Load (TMDL) for pollutants of concern. Human activity impacting water quality will then be controlled via other regulatory means in order to achieve the TMDL targets.

Water designated for human consumption as drinking water may be subject to specific drinking water quality standards. In the United States, for example, such standards have been developed under the Safe Drinking Water Act, are mandatory, and are enforced via a comprehensive monitoring and correction program.

Effluent Limitations

Effluent limitations are legal requirements governing the discharge of pollutants into water. Such standards set quantitative limits on the permissible amount of specific water pollutants that may be released from specific sources over specific timeframes. They are generally designed to achieve water quality standards for the receiving waterbody.

Numerous methods exist for determining appropriate limitations...

For example, many of these approaches are used in the United States. The law requires the United States Environmental Protection Agency (responsible for water quality regulation at a national level under the U.S. Clean Water Act to develop "effluent guidelines" - national industry-specific effluent limitations based on the performance of existing control technologies, including "best conventional pollutant control technology," "best practicable control technology," and "best available technology economically achievable."

Around the World

International Law

Marine and ship pollution are serious threats to the world's oceans. The London Convention limits ocean dumping from vessels, aircraft and platforms. MARPOL 73/78 also governs ship pollution.

Canada

The Canada Water Act is the principal federal law protecting Canadian waters. It is administered by Environment Canada.

Guidelines for Canadian Drinking Water Quality contains federal drinking water standards. It is administered by Health Canada.

Water Resources Law

Water resources law (in some jurisdictions, shortened to "water law") is the field of law dealing with the ownership, control, and use of water as a resource. It is most closely related to property law, and is older than and distinct from laws governing water quality.

Waters Subject to Regulation

Water is ubiquitous and does not respect political boundaries. Water resources laws may apply to any portion of the hydrosphere over which claims may be made to appropriate or maintain the water to serve some purpose. Such waters include, but are not limited to:

- Surface waters - lakes, rivers, streams, oceans, and wetlands;

- Surface runoff—generally water that flows across the land from rain, floodwaters, and snowmelt before those waters reach watercourses, lakes, wetlands, or oceans;

- Groundwater, particularly water present in aquifers.

History

The history of people's relation to water illustrates varied approaches to the management of water resources. "Lipit Ishtar and Ur Nammu both contain water provisions, pre-date Hammurabi by at least 250 years, and clearly provide the normative underpinnings on which the Hammurabi Code was constructed." The Code of Hammurabi was one of the earliest written laws to deal with water issues, and this Code included the administration of water use. The Code was developed about 3,800 years ago by King Hammurabi of Babylonia.

Difficulties of Water Rights

Water has unique features that make it difficult to regulate using laws designed mainly for land. Water is mobile, its supply varies by year and season as well as location, and it can be used simultaneously by many users. As with property (land) law, water rights can be described as a "bundle of sticks" containing multiple, separable activities that can have varying levels of regulation. For instance, some uses of water divert it from its natural course but return most or all of it (e.g. hydroelectric plants), while others consume much of what they take (ice, agriculture), and still others use water without diverting it at all (e.g. boating). Each type of activity has its own needs and can in theory be regulated separately. There are several types of conflict likely to arise: absolute shortages; shortages in a particular time or place; diversions of water that reduce the flow available to others; pollutants or other changes (such as temperature or turbidity) that render water unfit for others' use; and the need to maintain "in-stream flows" of water to protect the natural ecosystem.

One theory of history, put forward in the influential book *Oriental Despotism*, holds that many empires were organized around a central authority that controlled a population through monopolizing the water supply. Such a hydraulic empire creates the potential for despotism, and serves as a cautionary tale for designing water regulations.

Water law involves controversy in some parts of the world where a growing population faces increasing competition over a limited natural supply. Disputes over rivers, lakes and underground aquifers cross national borders. Although water law is still regulated mainly by individual countries, there are international sets of proposed rules such as the Helsinki Rules on the Uses of the Waters of International Rivers and the Hague Declaration on Water Security in the 21st Century.

Long-term issues in water law include the possible effects of global warming on rainfall patterns and evaporation; the availability and cost of desalination technology; the control of pollution, and the growth of aquaculture.

Legal Models

The legal right to use a designated water supply is known as a water right. There are two major models used for water rights. The first is riparian rights, where the owner of the adjacent land has the right to the water in the stream. The other major model is the prior appropriations model, the first party to make use of a water supply has the first rights to it, regardless of whether the property is near the water source. Riparian systems are generally more common in areas where water is plentiful, while appropriations systems are more common in dry climates. As water resource law is complex, many areas have some combination of the two approaches.

Water Law by Country

International Law

The right to water to satisfy basic human needs for personal and domestic uses has been protected under international human rights law. When incorporated in national legal frameworks, this right is articulated to other water rights within the broader body of water law. The human right to water has been recognized in international law through a wide range of international documents, including international human rights treaties, declarations and other standards.

The human right to water places the main responsibilities upon governments to ensure that people can enjoy "sufficient, safe, accessible and affordable water, without discrimination". Most especially, governments are expected to take reasonable steps to avoid a contaminated water supply and to ensure there are no water access distinctions amongst citizens. Today all States have at least ratified one human rights convention which explicitly or implicitly recognizes the right, and they all have signed at least one political declaration recognizing this right.

Canada

Under the *Constitution Act, 1867*, jurisdiction over waterways is divided between the federal and provincial governments. Federal jurisdiction is derived from the powers to regulate navigation and shipping, fisheries, and the governing of the northern territories, which has resulted in the passage of:

- the *Fisheries Act*,
- the *Navigable Waters Protection Act*,
- the *Arctic Waters Pollution Prevention Act*, and
- the *Oceans Act*.

Provincial jurisdiction is derived from the powers over property and civil rights, matters of a local and private nature, and management of Crown lands. In Ontario, Quebec and other provinces, the beds of all navigable waters are vested in the Crown, in contrast to English law. All provincial governments also govern water quality through laws on environmental protection and drinking water, such as the *Clean Water Act* in Ontario.

Australia

Water law in Australia varies with each state.

Tasmania

A newly formed Tasmanian Water Corporation has compulsorily acquired all drinking water supply infrastructure without payment and does not have direct accountability

Water Law in the United States

In the United States there are complex legal systems for allocating water rights that vary by region. These varying systems exist for both historical and geographic reasons. Water law encompasses a broad array of subjects or categories designed to provide a framework to resolve disputes and policy issues relating to water:

- Public waters, including tidal waters and navigable waterways.

- Other surface waters—generally water that flows across non-public land from rain, flood-waters, and snowmelt before those waters reach public watercourses.

- Groundwater, sometimes called subterranean, percolating, or underground water

- Public regulation of waters, including flood control, environmental regulation—state and federal, public health regulation and regulation of fisheries

- Related to all of the above is interplay of public and private rights in water, which draws on aspects of eminent domain law and the federal commerce clause powers

- Water project law: the highly developed law regarding the formation, operation, and finance of public and quasi-public entities which operate local public works of flood control, navigation control, irrigation, and avoidance of environmental degradation

- Treaty Rights of Native Americans

The law governing these topics comes from all layers of law. Some derives from common law principles which have developed over centuries, and which evolve as the nature of disputes presented to courts change. For example, the judicial approach to landowner rights to divert surface waters has changed significantly in the last century as public attitudes about land and water have evolved. Some derives from state statutory law. Some derives from the original public grants of land to the States and from the documents of their origination. Some derives from state, federal and local regulation of waters through zoning, public health and other regulation. Non-federally recognized Indian tribes do not have water rights.

Water Law in the European Union

For countries within the European Union, water-related directives are important for water resource management and environmental and water quality standards. Key directives include the Urban Waste Water Directive 1992 (requiring most towns and cities to treat their wastewater to specified standards), and the Water Framework Directive 2000/60/EC, which requires water resource plans based on river basins, including public participation based on Aarhus Convention principles.

Water Resource Policy

Water resource policy encompasses the policy-making processes that affect the collection, preparation, use and disposal of water to support human uses and protect environmental quality.

Water policy addresses provision, use, disposal and sustainability decisions. Provision includes identification, access, preparation for use and distribution. Uses include direct human consumption, agriculture, industry and ecosystem protection. Policy must set the rules for how water is allocated to the different uses. Disposal involves wastewater treatment and stormwater/flood management. Sustainability addresses issues such as aquifer depletion, reservoir management and mineral buildup.

"Supply isn't just about water production, it is also about distribution infrastructure."

A second dimension of issues addresses how policies are created, executed and amended. Since water resources often cross political boundaries, water policies must often be negotiated among multiple political entities (nations, states, etc.) Commentators such as Halcrow project resource wars as demand continues to increase.

Policy makers typically adopt a set of best management practices BMPs to govern water management. BMPs cover everything from dam construction to wastewater treatment protocols.

Water resource policies may encompass

"regions, catchments, shared or transboundary water resources, and inter-basin transfers. Policy leads management practices, but best management practices are identified, evaluated, modified and disseminated by policy making bodies."

Water resource policy issues are receiving increased attention as water shortages are believed to be at crisis levels in some regions. These regional crises have the potential worldwide implications.

Organizations such as the Global Water Policy Project have sprung up to promote awareness and prod governments and NGOs into heightened awareness of the problems.

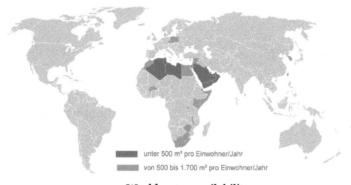

World water availability

Global Water Resource Policy Objectives

According to the World Water Assessment Programme, a UN-sanctioned Task Force, the objectives for global water resource policies include developing a standardized method for monitoring water sector progress and performance, improving reporting and identifying priority actions. In all nations conflict between users are expected to intensify, complicating policy-making.

Institutional Participants

Multilateral

The 1977 Mar del Plata United Nations Conference on Water was the first intergovernmental water conference, leading to the 1980 Declaration of the International Drinking Water Supply and Sanitation Decade by the UN General Assembly.

The United Nations Environmental Program hosts water resource policy-making agencies and disseminating BMPs worldwide. This role has been enhanced by various policy directives and other initiatives:

- UN General Assembly Resolution 3436 (XXX) Agenda 21
- 1997 Nairobi Declaration on the Role and Mandate of UNEP and
- 2000 Malmö Ministerial Declaration adopted at the First Global Ministerial Environment Forum.
- 2002 Earth Summit 2015 safe drinking water targets.
- 2007 World Bank report series on Environment and Development that in 2009 reported on Environmental Flows in Water Resources Policies, Plans, and Projects

Bilateral

Treaties between nations may enumerate policies, rights and responsibilities. For instance, a treaty between Poland and Germany, "An Agreement to establish cooperation on water resources management" provides:

- supply of drinking water of good quality,
- protection of surface water,
- supply of water to agriculture,
- fight against water pollution.

The Permanent Court of International Justice adjudicates disputes between nations, including water rights litigation.

NGOs

Some non-governmental organizations have consultative status at the UN. One such group is the World Water Council, an "international multi-stakeholder platform" established in 1996 to act "at all levels, including the highest decision-making level...[in] protection, development, planning, management and use of water in all its dimensions...for the benefit of all life on earth." It was an outgrowth of the 1992 UN Conference on Environment and Development in Dublin and at the Rio de Janeiro Earth Summit. The Council is based in Marseilles. Its multi-stakeholder basis as due to the fact that "authority for managing the world's fresh water resources is fragmented amongst the world's nations, hundreds of thousands of local governments, and countless non-governmental and private organizations, as well as a large number of international bodies."

In 1994, the International Water Resources Association (IWRA) organized a special session on the topic in its Eighth World Water Congress held in Cairo in November 1994, leading to creation of the World Water Council.

Business Water Resource Policy Initiatives

The World Business Council for Sustainable Development engages stakeholders in H2OScenarios that consider various alternative policies and their effects.

In June 2011 in Geneva, the Future of Water Virtual Conference addressed water resource sustainability. Issues raised included: water infrastructure monitoring, global water security, potential resource wars, interaction between water, energy, food and economic activity, the "true value" of "distribution portions of available water" and a putative "investment gap" in water infrastructure. It was asserted that climate change will affect scarcity of water but the water security presentation emphasized that a combined effect with population growth "could be devastating". Identified corporate water related risks include physical supply, regulatory and product reputation.

This forum indicated policy concerns with:

- trade barriers

- price supports

- treatment of water as a free good creates underpricing of 98% of water

- need to intensify debate

- need to harmonize public/private sectors

Structural Constraints on Policy Makers

Policies are implemented by organizational entities created by government exercise of state power. However, all such entities are subject to constraints upon their autonomy.

Jurisdictional Issues

Subject matter and geographic jurisdiction are distinguishable.

The jurisdiction of any water agency is limited by political boundaries and by enabling legislation.

In some cases, limits target specific types of uses (wilderness, agricultural, urban-residential, urban-commercial, etc.)

A second part of jurisdictional limitation governs the subject matter that the agency controls, such as flood control, water supply and sanitation, etc.

In many locations, agencies may face unclear or overlapping authority, increasing conflicts and delaying conflict resolution. For instance, recent changes in California law intended to reduce air quality problems from shipping have been interfered with by Federal legal changes intended to reduce the cost of shipping.

California Water Regulatory Bodies

- Coastal Commission

- Coastal Conservancy

- Department of Fish & Game

- Department of Water Resources

- Environmental Resources Evaluation System (CERES)

- Ocean and Coastal Environmental Access Network (OCEAN)

- Resources Agency Wetlands Information System

- State Water Resources Control Board

- Public Health Departments

- Water districts

Typical Information Access Issue

As reported by the non-partisan Civil Society Institute, a 2005 US Congressional study on water supply was suppressed and became the target of a Freedom of Information Act (FOIA) litigation.

Issues

Flood Control

Water can produce a natural disaster in the form of tsunamis, hurricanes, rogue waves and storm surge. Land-based floods can originate from bursting dams, rivers overflowing their banks or levee failure.

Multi-jurisdictional Issues

One jurisdiction's projects may cause problems in other jurisdictions. For instance, Monterey County, California controls a body of water that acts as a reservoir for San Luis Obispo County. The specific responsibilities for managing the resource must therefore be negotiated.

Freshwater

Surface and Groundwater

Surface water and groundwater can be studied and managed as separate resources as a single resource in multiple forms. Jurisdictions typically distinguish three recognized groundwater classifications: subterranean streams, underflow of surface waters, and percolating ground-water.

Constituencies

Drinking water and water for utilitarian uses such as washing, crop cultivation and manufacture is competed for by various constituencies:

- Residential

- Agriculture. "Many rural people practice subsistence rain fed agriculture as a basic liveli-hood strategy, and as such are vulnerable to the effects of drought or flood that can dimin-ish or destroy a harvest. "

- Construction

- Industrial

- Municipal or institutional activities

Seawater

Seawater resources are important for ethical-aesthetic reasons, recreation, tourism, maintenance of fisheries. The sea is a venue for shipping and for oil and mineral extraction that creates a need for regulatory policy. A variety of issues confronts policy makers.

Pollution

Ballast water, fuel/oil leaks and trash originating from ships foul harbors, reefs and estuaries. Bal-last water may contain toxins, invasive plants, animals, viruses, and bacteria.

Oil rigs and undersea mineral extraction can create problems that affect shorelines, marine life, fisheries and human safety. Decommissioning of such operations has another set of issues. Rigs-to-reefs is a proposal for using obsolete oil rigs as substrate for coral reefs that has failed to reach consensus.

Surface Water (Runoff) and Wastewater Discharge

Regulatory bodies address piped waste water discharges to surface water that include riparian and ocean ecosystems. These review bodies are charged with protecting wilderness ecology, wild-life habitat, drinking water, agricultural irrigation and fisheries. Stormwater discharge can car-ry fertilizer residue and bacterial contamination from domestic and wild animals. They have the authority to make orders which are binding upon private actors such as international corpora-tions and do not hesitate to exercise the police powers of the state. Water agencies have statutory mandate which in many hurisdictions is resilient to pressure from constituents and lawmakers in which they on occasion stand their ground despite heated opposition from agricultural interests On the other hand, the Boards enjoy strong support from environmental concerns such as Green-peace,Heal the Ocean and Channelkeepers.

Water quality issues or sanitation concerns reuse or water recycling and pollution control which in turn breaks out into stormwater and wastewater.

Stormwater Runoff

Surface runoff is water that flows when heavy rains do not infiltrate soil; excess water from rain, meltwater, or other sources flowing over the land. This is a major component of the water cycle. Runoff that occurs on surfaces before reaching a channel is also called a nonpoint source. When runoff flows along the ground, it can pick up soil contaminants including, but not limited to petroleum, pesticides, or fertilizers that become discharge or nonpoint source pollution.

Wastewater

Wastewater is water that has been discharged from human use. The primary discharges flow from the following sources:

- residences

- commercial properties

- industry

- agriculture

Sewage is technically wastewater contaminated with fecal and similar animal waste byproducts, but is frequently used as a synonym for wastewater. Origination includes cesspool and sewage outfall pipes.

Water treatment is subject to the same overlapping jurisdictional constraints which affect other aspects of water policy. For instance, levels of chloramines with their resulting toxic trihalomethane by-product are subject to Federal guidelines even though water management implementing those policy constraints are carried out by local water boards.

Xerochore

XEROCHORE is an Exercise to Assess Research Needs and Policy Choices in Areas of Drought founded by European Commission under the FP7-Theme 6, Environment (Including Climate Change), and it is aimed at assisting in the development of a European Drought Policy in accordance with the EU-Water Framework Directive (EU-WFD.

Background and Objectives

In recent years large parts of Europe suffered from extreme drought, a phenomenon that may become more frequent and more severe as a consequence of climate change. If not addressed properly, the socio-economic and environmental impacts of droughts will be huge.

To inform European Drought and Climate Adaptation policies, the Xerochore assesses the existing knowledge about the physical causes, intensity, spatial and temporal patterns of droughts, and their social, environmental and economic impacts. Furthermore, it develops a catalogue of demand-and supply-side policies to manage droughts.

To this end a series of expert workshops, roundtable discussions and stakeholders consultation will be organised. A Core Group will synthesise the collected knowledge and produce Guidance Documents for drought management.

The Xerochore SA objectives are:

- To synthesize knowledge on past, current and future drought events;

- To compile a roadmap that provides a short to long term vision on research needs and steps forward towards supporting the implementation of drought management plans;

- To provide information on possible impacts of droughts and guidance for stakeholders in the area of planning, implementation and scenarios;

- To further extend and develop the drought network established as part of the European Drought Centre (EDC) to assess the international (inside and outside of Europe) state of the art in research related to droughts;

- To link up with non-EU experts (through network partners and members of the External Advisory Board) to ensure that the synthesis, roadmap and guidance to stakeholders includes background from other leading continents (e.g. USA, Canada, Australia) and to encourage them to become member of the extended EDC;

- To initiate a long lasting platform beyond this project through the network that communicates drought related research and policy making within the research community, water managers, policy makers and the wider public.

Organization of Work

The Xerochore SA is based on the involvement of 11 Partners (Core Partners) and a wider group of already identified partners as Network Partners (contributing to the review and synthesis). Also being open to more experts on a voluntary basis, the area addressed benefits from broadening the information collection as much as possible, by keeping the effort for coordinating the synthesis development in reasonable limits. The Xerochore SA is related to five main areas corresponding to 5 work packages (WP1-5), addressing sectoral key topics for the final synthesis. A sixth work package deals with the management of the SA (WP6). The five main areas (work packages) to be covered are:

WP1: Natural system (climate and hydrology), including physical causes, spatial and temporal development of historic droughts, anthropogenic influences, incl also impact of climate change on drought, development of basic elements for early warning systems, forecasting of drought;

WP2: Economic and social impacts of droughts and of research actions taken to mitigate and to adapt to the effects of droughts, incl. providing guidance for stakeholders;

WP3: Environmental impacts of droughts and of research actions taken to mitigate and to adapt to the effects of droughts, incl. providing guidance for stakeholders;

WP4: An appraisal of existing drought-associated policies and the research needs to develop an EU drought policy (Pan-European scale);

WP5: Implications of drought policies for integrated water management at the regional and national level.

The outcome of the WP's 1-3 provide then a possibility to integrate drought management both to its drivers (causes) and impacts as well as to possible consequences of drought management (WP5). The link of the drought management to WP4 (Drought policy) is dual, first to impose urgent fields of action in policy making, but considering at the same time constraints and restrictions as discussed under WP5.

Partners

The XEROCHORE Support Action are implemented by a Consortium of 11 Core Partners. The Consortium composition ensures interdisciplinary and multiplicity of skills in order to allow the achievement of the multifaceted project objectives and ensure a successful implementation of the outreach strategy by reaching all the different target groups addressed by the project activities. The project network consists of over 80 organizations including research institutes, universities, ministries, water management organizations, stakeholders, consultants, international organizations and programmes. The XEROCHORE Consortium, experienced in the different fields that lead toward a European Drought Policy, is well balanced as four main different partner types are represented:

International Recognised Research Centres (FEEM, WMC, NERC, CEMACREF, JRC): The involved International Research Centres contribute to long-term interactions with national and international meteorological services and will support the XEROCHORE roadmap on applied projects and consulting. These research centres enable sufficient consideration of practical needs and requirements for private sector markets, within the roadmap towards EU drought policy.

Academic Institutions (WU, UiO, NTUA): The involvement of these type of institutions together with similar Network Partners will ensure that the very first outcome of ongoing international academic research on different drought issues that is carried out by PhD students and postdocs (e.g. Aquastress, WATCH) will be integrated in the round table discussions that eventually will lead to the XEROCHORE roadmap for future research.

National Authorities (MATTM, MMA): The involved National Authorities contribute to the XEROCHORE project by interactions with national and international hydrometeorological services and will support the roadmap in consideration of institutional links and requirements from national authorities and governmental organisations.

International Organisation (JRC, IUCN): The involvement of International Organisations contributes to the embedding of international NGO's and support to policy making at EU and international level in consideration of environmental aspects and sustainable development.

Water Right

Water right in water law refers to the right of a user to use water from a water source, e.g., a river, stream, pond or source of groundwater. In areas with plentiful water and few users, such systems are generally not complicated or contentious. In other areas, especially arid areas where irrigation is practiced, such systems are often the source of conflict, both legal and physical. Some systems treat surface water and ground water in the same manner, while others use different principles for each.

Types of Water Rights

Fundamental differences exist between the nature and source of water rights in different countries. Generally, water rights are based on the water law that applies in a particular country and, at their most basic, are classified as land-based or use-based rights.

Some countries allow their subdivisions to establish independent water laws. For example, each state and territory of the United States has its own set of water laws that establish water rights that may be land-based, use-based, or both.

Land-based

Some water rights are based on land ownership or possession. For example, many common law jurisdictions recognize riparian rights, which are protected by property law. Riparian rights state that only the owner of the banks of the water source have a right to the 'undiminished, unaltered flow' of the water.

Finnish Water Rights

In Finland, waterbodies are generally privately owned, which is not the case in most EU countries, but Finland also applies the Roman law principle of *aqua profluens* (flowing water), according to which the freely flowing water in waterbodies cannot be owned or possessed. This means that the owners of waterbodies cannot prohibit diversion of water for agricultural, industrial, municipal, or domestic use according to the provisions of the Finnish Water Law and cannot prohibit use of the waterbodies for recreational purposes.

Community-based Rights

In some jurisdictions water rights are granted directly to communities and water is reserved to provide sufficient capacity for the future growth of that particular community. For example, California provides communities and other water users within watersheds senior status over appropriative (use-based) water rights solely because they are located where the water originates and naturally flows. A second example of community-based water rights is pueblo water rights. As recognized by California, pueblo water rights are grants to individual settlements (i.e. pueblos) over all streams and rivers flowing through the city and to all groundwater aquifers underlying that particular city. The pueblo's claim expands with the needs of the city and may be used to supply the needs of areas that are later annexed to the city. While California recognizes pueblo water rights, pueblo water rights are controversial. Some modern scholars and courts argue that the pueblo water rights doctrine lacks a historical basis in Spanish or Mexican water law.

Use-based Rights

Use-based rights do not relate to land and instead rely on whether the water user has legal access to the water source. As a general rule, use-based rights are usufructuary, fully transferable to anyone. Under common law use-based rights only apply to navigable-in-fact waterways in which there is a presumptive easement, often referred to as a navigable servitude.

Appropriation

Appropriative water rights are the most common use-based water rights in the United States and are most commonly found in the western states where water is scarcest. "The appropriation doctrine confers upon one who actually diverts and uses water the right to do so provided that the water is used for reasonable and beneficial uses," regardless of whether that person owns land contiguous to the watercourse. "[A]s between appropriators, the rule of priority is 'first in time, first in right.'" The modern system of prior appropriation water rights is characterized by five principles:

1. Exclusive right is given to the original appropriator, and all following privileges are conditional upon precedent rights.

2. All privileges are conditional upon beneficial use.

3. Water may be used on riparian lands or non-riparian lands (i.e. water may be used on the land next to the water source, or on land removed from the water source)

4. Diversion is permitted, regardless of the shrinkage of the river or stream.

5. The privilege may be lost through non-use.

Beneficial use is defined as agricultural, industrial, or urban use. Environmental uses, such as maintaining body of water and the wildlife that use it, were not initially regarded as beneficial uses in some states but have been accepted in some areas. Every water right is parameterized by an annual yield and an appropriation date. When a water right is sold, it maintains its original appropriation date.

Appropriative water rights are not always applied exclusively, as demonstrated by California which recognizes several different forms of water rights concurrently just for surface water. It recognizes a separate set of water rights for groundwater.

In-stream Water Rights

In-stream water rights are rights that only apply to water in a stream, and cannot be diverted for usage. These rights are most commonly used to protect endangered species or to bolster the number of a threatened aquatic species.

History of Water Rights

In Roman times, the law was that people could obtain temporary usufructuary rights for running water. These rights were independent of land ownership, and lasted as long as use continued. Under Roman law, no land was "owned" by citizens, it was all owned by the "republic" and controlled by politicians.

In Medieval times, the common law of the day treated all freshwater streams as static, meaning landowners owned parts of rivers, with full accompanying rights. Landowners could also seek damages for loss of water diverted upstream. Non-landowners did not have use rights, except by obtaining a prescription.

Over time, rights evolved from being land based to use based, allowing non-landowners to hold enforceable rights. A reasonable use rule evolved in some countries.

Water Rights in the United States

In the United States, there are two divergent systems for determining water rights. Riparian water rights (derived from English common law) are common in the east and prior appropriation water rights (developed in Colorado and California) are common in the west. Each state has its own variations on these basic principles, as informed by custom, culture, geography, legislation and case law. Californian law, for example, includes elements of both systems. In general, a water right is established by obtaining an authorization from the state in the form of a water right permit. A legal right is formally consummated, or perfected, by exercising the water right permit and using the water for a beneficial purpose.

Under the prior appropriation doctrine, water rights are "first in time, first in right". That is, the older, or senior, water right may operate to the exclusion of junior water rights. The concept of "priority date" is significant. The priority date is generally associated with the date that water was first put to beneficial use, or the date that a successful application for a water right was submitted, and indicates the relative status of seniority among competing users. Older rights are senior. More recent rights are junior.

Water rights are generally established pursuant to state law, but there are exceptions, most notably, the concept of federal reserved water rights. Federal reserved water rights are superimposed over state water systems and exhibit several unique characteristics. They are superior to the rights of subsequent water users, cannot be lost through non-use, and are immune to state-specific standards such as the "beneficial" and "reasonable" use doctrines. Reserved water rights are rights that are established when the federal government reserves land for a specific federal purpose. Courts have held that there is an implied water right to provide the minimum amount of water necessary to effectuate the primary purposes of the reservation . Examples of reservations include Indian reservations, national wildlife refuges, federal forests and military bases. The federal government has also successfully invoked federal reserved water rights for several entities administered by the National Park Service such as Rocky Mountain and Yellowstone National Parks as well as the Red Wild and Scenic River in New Mexico.

Proceedings to determine the relative priority of claims to water rights are known as adjudications. Through Congress's passage of the McCarren amendment, the federal government has consented to having its claims adjudicated in state courts.

All states offer mechanisms for changing how a water right is exercised, e.g., amending the point of diversion or withdrawal, the place of use and the purpose of use. In reviewing such requests, the state must guard against the impairment of other water rights, the enlargement of the water right and injury to the public interest.

Water rights generally emerge from a person's ownership of the land bordering the banks of a watercourse or from a person's actual use of a watercourse. Water rights are conferred and regulated by judge-made common law, state and federal legislative bodies, and other government departments. Water rights can also be created by contract, as when one person transfers his water rights to another.

In the eighteenth century, regulation of water was primarily governed by custom and practice. As the U.S. population expanded over the next two centuries, however, and the use of water for agrarian and domestic purposes increased, water became viewed as a finite and frequently scarce resource. As a result, laws were passed to establish guidelines for the fair distribution of this resource. Courts began developing common-law doctrines to accommodate landowners who asserted competing claims over a body of water. These doctrines govern three areas: riparian rights, surface water rights, and underground water rights.

An owner or possessor of land that abuts a natural stream, river, pond, or lake is called a riparian owner or proprietor. The law gives riparian owners certain rights to water that are incident to possession of the adjacent land. Depending on the jurisdiction in which a watercourse is located, riparian rights generally fall into one of three categories.

First, riparian owners may be entitled to the "natural flow" of a watercourse. Under the natural flow doctrine, riparian owners have a right to enjoy the natural condition of a watercourse, undiminished in quantity or quality by other riparian owners. Every riparian owner enjoys this right to the same extent and degree, and each such owner maintains a qualified right to use the water for domestic purposes, such as drinking and bathing.

However, this qualified right does not entitle riparian owners to transport water away from the land abutting the watercourse. Nor does it permit riparian owners to use the water for most irrigation projects or commercial enterprises. Sprinkling gardens and watering animals are normally considered permissible uses under the natural flow doctrine of riparian rights.

Second, riparian owners may be entitled to the "reasonable use" of a watercourse. States that recognize the reasonable use doctrine found the natural flow doctrine too restrictive. During the industrial revolution of the nineteenth century, some U.S. courts applied the natural flow doctrine to prohibit riparian owners from detaining or diverting a watercourse for commercial development, such as manufacturing and milling, because such development impermissibly altered the water's original condition.

In replacing the natural flow doctrine, a majority of jurisdictions in the United States now permit riparian owners to make any reasonable use of water that does not unduly interfere with the competing rights and interests of other riparian owners. Unlike the natural flow doctrine, which seeks to preserve water in its original condition, the reasonable use doctrine facilitates domestic and commercial endeavors that are carried out in a productive and reasonable manner.

When two riparian owners assert competing claims over the exercise of certain water rights, courts applying the reasonable use doctrine generally attempt to measure the economic value of the water rights to each owner. Courts also try to evaluate the prospective value to society that would result from a riparian owner's proposed use, as well as its probable costs. No single factor is decisive in a court's analysis.

Third, riparian owners may be entitled to the "prior appropriation" of a watercourse. Where the reasonable use doctrine requires courts to balance the competing interests of riparian owners, the doctrine of prior appropriation initially grants a superior legal right to the first riparian owner who makes a beneficial use of a watercourse. The prior appropriation doctrine is applied in most arid western states, including Arizona, Colorado, Idaho, Montana, Nevada, New Mexico, Utah, and Wyoming and requires the riparian owner to demonstrate that she is using the water in an economically efficient manner. Consequently, the rights of a riparian owner under the prior appropriation doctrine are always subject to the rights of other riparian owners who can demonstrate a more economically efficient use.

Under any of the three doctrines, the interests of riparian owners are limited by the constitutional authority of the state and federal governments. The Commerce Clause of the U.S. Constitution gives Congress the power to regulate Navigable Waters, a power that Congress has exercised in a variety of ways, including the construction of dams. In those instances where Congress does not exercise its power under the Commerce Clause, states retain authority under their own constitutions to regulate waterways for the public good.

However, the eminent domain clause of the Fifth Amendment to the U.S. Constitution limits the power of state and federal governments to impinge on the riparian rights of landowners by prohibiting the enactment of any laws or regulations that amount to a "taking" of private property. Laws and regulations that completely deprive a riparian owner of legally cognizable water rights constitute an illegal governmental taking of private property for Fifth Amendment purposes. The Fifth Amendment requires the government to pay the victims of takings an amount equal to the fair market value of the water rights.

Some litigation arises not from the manner in which neighboring owners appropriate water but from the manner in which they get rid of it. The disposal of surface waters, which consist of drainage from rain, springs, and melting snow, is typically the source of such litigation. This type of water gathers on the surface of the earth but never joins a stream, lake, or other well-defined body of water.

Litigation arises when one owner drains excess surface water onto neighboring property. Individuals who own elevated property may precipitate a dispute by accelerating the force or quantity of surface water running downhill, and individuals who own property on a lower level may rankle their neighbors by backing up surface water through damming and filling. Courts are split on how to resolve such disputes.

Some courts apply the common-law rule that allows landowners to use any method of surface water removal they choose without liability for flooding that may result to nearby property. Application of this rule generally rewards assertive and clever landowners and does not discourage neighbors from engaging in petty or vindictive squabbles over surface water removal.

Other courts apply the civil-law rule, which stems from Louisiana, a civil-law jurisdiction. This rule imposes Strict Liability for any damage caused by a landowner who interrupts or alters the natural flow of water. The civil-law rule encourages neighbors to let nature take its course and live with the consequences that may follow from excessive accumulation of standing surface water.

Over the last quarter century many courts have begun applying the reasonable use rule to surface water disputes. This rule enables landowners to make reasonable alterations to their land for drainage purposes as long as the alteration does not unduly interfere with a neighbor's right to do the same. In applying this rule, courts balance the neighbors' competing needs, the feasibility of more appropriate methods of drainage, and the comparative severity of injuries.

Surface water that seeps underground can also create conditions ripe for litigation. Sand, sod, gravel, and even rock are permeable substances in which natural springs may form and moisture can collect. Underground reservoirs can be tapped by artificial wells that are used in conjunction by commercial, municipal, and private parties. When an underground water supply is appreciably depleted by one party, other parties with an interest in the well may sue for damages.

As with surface water and riparian rights, three theories of underground water rights have evolved. The first theory, known as the absolute ownership theory, derives from English Law and affords landowners the right to withdraw as much underground water as they wish, for whatever purpose, requiring their neighbors to fend for themselves. Under the second theory, known as the American rule, landowners may withdraw as much underground water as they like as long as it is not done for a malicious purpose or in a wasteful manner. This theory is now applied in a majority of jurisdictions in the United States.

California has developed a third theory of underground water rights, known as the correlative theory. The correlative theory provides each landowner with an equal right to use underground water for a beneficial purpose. But landowners are not given the prerogative to seriously deplete a neighbor's water supply. In the event of water shortage, courts may apportion an underground supply among landowners. Many states facing acute or chronic shortages have adopted the correlative theory of under-ground water rights.

Water rights can also be affected by the natural avulsion or accretion of lands underlying or bordering a watercourse. Avulsions are marked by a sudden and violent change to the bed or course of a stream or river, causing a measurable loss or addition to land. Accretions are marked by the natural erosion of soil on one side of a watercourse and the gradual addition of soil to the other side. The extended shoreline made by sedimentary deposits is called an alluvion. Water rights are not altered by avulsions. However, any accretions of soil enure to the benefit of the landowner whose holdings have increased by the alluvion addition.

Although water covers more than two-thirds of the earth's surface, U.S. law treats water as a limited resource that is in great demand. The manner in which this demand is satisfied varies according to the jurisdiction in which a water supply is located. In some jurisdictions the most productive use is rewarded, whereas in other jurisdictions the first use is protected. Several jurisdictions are dissatisfied with both approaches and allow a water supply to be reasonably appropriated by all interested parties. Each approach has its weaknesses, and jurisdictions will continue experimenting with established legal doctrines to better accommodate the supply and demand of water rights.

Resolution of Interstate Water Conflicts

Because water bodies may cross political and jurisdictional boundaries, conflicts may arise. In the United States, three basic approaches are used to settle such conflicts: 1) Litigation before the Supreme Court of the United States; 2) Legislative resolution by the Congress of the United States; and 3) Negotiation and ratification of interstate compacts between states. In the western United States, for example, the 1922 Colorado River Compact divides the Colorado River basin into two areas, the Upper Division (comprising Colorado, New Mexico, Utah and Wyoming) and the Lower Division (Nevada, Arizona and California). A comprehensive review of existing interstate water compacts has been published by the Model Interstate Water Compact Project at the University of New Mexico School of Law's Utton Transboundary Resource Center.

Limitations on Water Rights

In California, courts have held that appropriation water rights may be limited under the public trust doctrine, a common law principle recognized by some courts, which holds that the public has access rights upon navigable waters and that navigable waters are held in trust for the use of the people. The public trust doctrine was invoked by the California Supreme Court in a case restricting the amount of water Los Angeles could divert from tributaries of Mono Lake. The case was filed by the Audubon Society and the Mono Lake Committee.

In the United States, navigable waters are subject to the commerce clause of the U.S. Constitution. The commerce clause provides the federal government the ability to restrict state issued water rights via, for example, the enforcement of water quality standards via the Federal Water Pollution Control Act (Clean Water Act), the Federal Power Act and the protection of endangered species via the Endangered Species Act.

References

- Sax, J. L., et al.. Legal Control of Water Resources: Cases and Materials (4th edition). Thomson/West (2006), ISBN 978-0-314-16314-1.

- Davis, Mackenzie; Masten, Susan (22 February 2013). Principles of Environmental Engineering Science: Third Edition. McGraw-Hill Higher Education. ISBN 978-0-07-749219-9.

- "Foia Lawsuit Targets U.S. Department Of Energy For Withholding 'Water Energy Roadmap' Ordered By Congress". Civilsocietyinstitute.org. Retrieved 2014-07-14.

Permissions

Index